THE TWENTY FIRST CENTURY Science

570

D0514422

Project Directors

Angela Hall   Emma Palmer

Robin Millar   Mary Whitehouse

Editors

Angela Hall   Anne Scott

Carol Usher

Authors

Cris Edgell   Neil Ingram   Carol Levick   Cliff Porter

Ann Fullick   Mike Kalvis   Nick Owens   Jacqueline Punter

THE UNIVERSITY *of* York

THE SALTERS' INSTITUTE

Nuffield Foundation

Contents

How to use this book

Welcome to Twenty First Century Science. This book has been specially written by a partnership between OCR, The University of York Science Education Group, The Nuffield Foundation Curriculum Programme, and Oxford University Press.

On these two pages you can see the types of page you will find in this book, and the features on them. Everything in the book is designed to provide you with the support you need to help you prepare for your examinations and achieve your best.

Module Openers

Why study?: This explains why what you are about to learn is useful to scientists.

Find out about: Every module starts with a short list of the things you'll be covering.

Ideas about Science: Here you can read about the key ideas about science covered in this module.

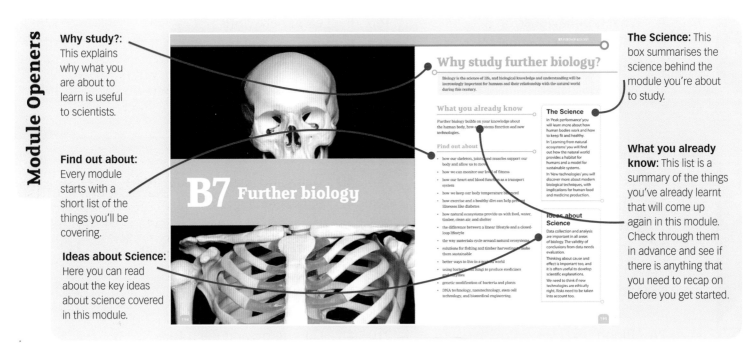

The Science: This box summarises the science behind the module you're about to study.

What you already know: This list is a summary of the things you've already learnt that will come up again in this module. Check through them in advance and see if there is anything that you need to recap on before you get started.

Main Pages

Find out about: For every part of the book you can see a list of the key points explored in that section.

Key words: The words in these boxes are the terms you need to understand for your exams. You can look for these words in the text in bold or check the glossary to see what they mean.

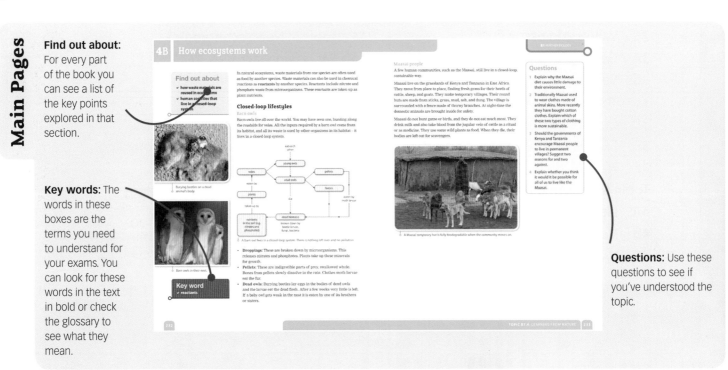

Questions: Use these questions to see if you've understood the topic.

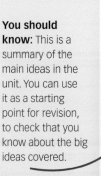

You should know: This is a summary of the main ideas in the unit. You can use it as a starting point for revision, to check that you know about the big ideas covered.

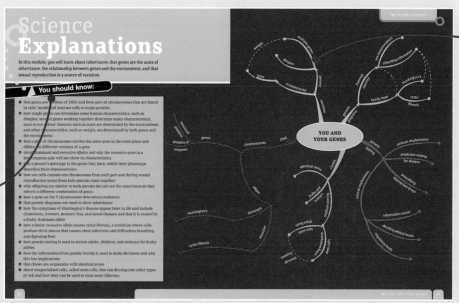

Visual summary: Another way to start revision is to use a visual summary, linking ideas together in groups so that you can see how one topic relates to another. You can use this page as a starting point for your own summary.

Ideas about Science: For every module this page summarises the ideas about science that you need to understand.

Review Questions: You can begin to prepare for your exams by using these questions to test how well you know the topics in this module.

Structure of assessment

Matching your course

What's in each module?

As you go through the book you should use the module opener pages to understand what you will be learning and why it is important. The table below gives an overview of which main topics each module includes.

B1
• What are genes and how do they affect the way that organisms develop? • Why can people look like their parents, brothers and sisters, but not be identical to them? • How can and should genetic information be used? How can we use our knowledge of genes to prevent disease? • How is a clone made?

B4
• How do chemical reactions take place in living things? • How do plants make food? • How do living organisms obtain energy?

B2
• How do our bodies resist infection? • What are vaccines and antibiotics and how do they work? • What factors increase the risk of heart disease? • How do our bodies keep a healthy water balance?

B5
• How do organisms develop? • How does an organism produce new cells? • How do genes control growth and development within the cell?

B3
• Systems in balance – how do different species depend on each other? • How has life on Earth evolved? • What is the importance of biodiversity?

B6
• How do animals respond to changes in their environment? • How is information passed through the nervous system? • What can we learn through conditioning? • How do humans develop more complex behaviour?

B7
• Peak performance – the skeletal system   • What can we learn from natural ecosystems? • Peak performance – circulation   • New technologies • Peak performance – energy balance

How do the modules fit together?

The modules in this book have been written to match the specification for GCSE Biology. In the diagram to the right you can see that the modules can also be used to study parts of GCSE Science and GCSE Additional Science.

	GCSE Biology	GCSE Chemistry	GCSE Physics
GCSE Science	B1	C1	P1
	B2	C2	P2
	B3	C3	P3
GCSE Additional Science	B4	C4	P4
	B5	C5	P5
	B6	C6	P6
	B7	C7	P7

GCSE Biology assessment

The content in the modules of this book matches the modules of the specification.

The diagram below shows you which modules are included in each exam paper. It also shows you how much of your final mark you will be working towards in each paper.

	Unit	Modules Tested			Percentage	Type	Time	Marks Available
Route 1	A161	B1	B2	B3	25%	Written Exam	1 h	60
	A162	B4	B5	B6	25%	Written Exam	1 h	60
	A163	B7			25%	Written Exam	1 h	60
	A164	Controlled Assessment			25%		4.5–6 h	64

Command words

The list below explains some of the common words you will see used in exam questions.

Calculate
Work out a number. You can use your calculator to help you. You may need to use an equation. The question will say if your working must be shown. (Hint: don't confuse with 'Estimate' or 'Predict'.)

Compare
Write about the similarities and differences between two things.

Describe
Write a detailed answer that covers what happens, when it happens, and where it happens. Talk about facts and characteristics. (Hint: don't confuse with 'Explain'.)

Discuss
Write about the issues related to a topic. You may need to talk about the opposing sides of a debate, and you may need to show the difference between ideas, opinions, and facts.

Estimate
Suggest an approximate (rough) value, without performing a full calculation or an accurate measurement. Don't just guess – use your knowledge of science to suggest a realistic value. (Hint: don't confuse with 'Calculate' and 'Predict'.)

Explain
Write a detailed answer that covers how and why a thing happens. Talk about mechanisms and reasons. (Hint: don't confuse with 'Describe'.)

Evaluate
You will be given some facts, data, or other kind of information. Write about the data or facts and provide your own conclusion or opinion on them.

Justify
Give some evidence or write down an explanation to tell the examiner why you gave an answer.

Outline
Give only the key facts of the topic. You may need to set out the steps of a procedure or process – make sure you write down the steps in the correct order.

Predict
Look at some data and suggest a realistic value or outcome. You may use a calculation to help. Don't guess – look at trends in the data and use your knowledge of science. (Hint: don't confuse with 'Calculate' or 'Estimate'.)

Show
Write down the details, steps, or calculations needed to prove an answer that you have given.

Suggest
Think about what you've learnt and apply it to a new situation or context. Use what you have learnt to suggest sensible answers to the question.

Write down
Give a short answer, without a supporting argument.

Top Tips

Always read exam questions carefully, even if you recognise the word used. Look at the information in the question and the number of answer lines to see how much detail the examiner is looking for.

You can use bullet points or a diagram if it helps your answer.

If a number needs units you should include them, unless the units are already given on the answer line.

Making sense of graphs

Reading the axes

Look at these two charts, which both provide data about daily energy use in several countries.

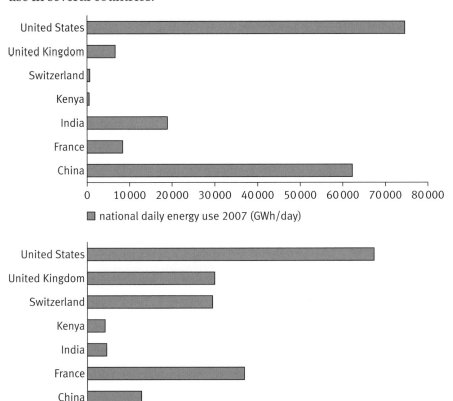

national daily energy use 2007 (GWh/day)

daily energy use per person (kWh per person/day)

Why are the charts so different if they both represent data about energy use?

One shows **energy use per person per day**, the other shows the **energy use per day by the whole country**.

First rule of reading graphs: read the axes and check the units.

Describing the relationship between variables

The pattern of points plotted on a graph shows whether two factors are related.

Look at this scatter graph.

Is there evidence that the two variables:
a are correlated?
b show a causal relationship?

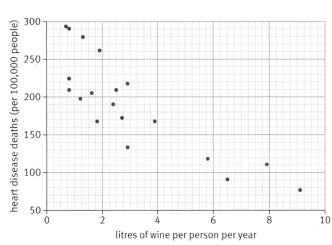

Graph showing that moderate wine drinking may reduce heart disease, from a study over 19 countries.

9

Rates of reaction

It is often useful to show changes in a variable over time. Enzyme reactions produce product over time, as shown in the graph on the right.

Q What happens to the amount of product being produced as this enzyme reaction carries on?

Q What happens to the speed of production of the product over time? Hint – look at the gradient of the line. This tells you how the amount of product produced in a given amount of time changes.

Q Sketch a graph of the rate of this reaction over the same time period.

Look at the graph on the right, which shows how the number of bacteria infecting a patient changes over time.

How many different gradients can you see?

There are three phases to the graph, each with a different gradient. So you should describe each phase, including **data** if possible:

- The number of bacteria **increases rapidly** for the first day until there are about **4.5 million** bacteria.
- For about **the next three days** the number remains steady at about 4.5 million.
- After the **fourth** day the number of bacteria declines to less than a **million** over the following **two to three days**.

Second rule of reading graphs: describe each distinct phase of the graph, and include ideas about the **gradient** and **data** including **units**.

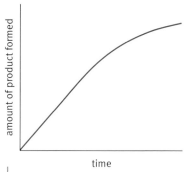

Graph showing product produced over time by an enzyme reaction.

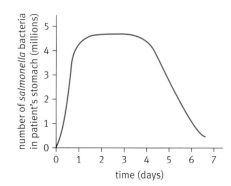

Frequency data

Frequency graphs or charts show the number of times a data value occurs. For example, if four students have a pulse rate of 86, then the data value 86 has a frequency of 4.

A large data set with lots of different values can be arranged into class intervals (or groups). Collecting data in class intervals can be done by tallying. It works well to have data arranged in five or six class intervals.

Class interval	Tally	Frequency
60–65	|	1
65–70	||||	4
70–75	⊦⊦⊦ ⊦⊦⊦ ||	12
75–80	⊦⊦⊦ |||	8
80–85	⊦⊦⊦	5
85–90	|	1
	Total	**31**

A data set of pulse rates from a class of 31 pupils tallied in a frequency table.

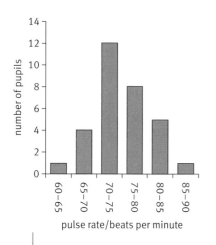

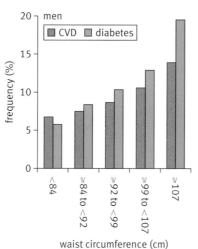

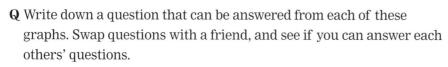

Frequency data can be shown in a bar chart. This bar chart shows the pulse rates of a class of 15-year-old pupils at rest.

Sometimes frequency graphs have % as the units on the y axis. CVD: cardiovascular disease.

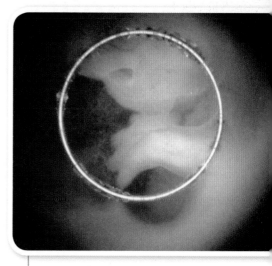

A human fetus at eight weeks. This photograph was taken from inside the mother's uterus. The fetus is about 2.5 cm long.

Q Write down a question that can be answered from each of these graphs. Swap questions with a friend, and see if you can answer each others' questions.

Range and mean

Statistics are used to describe data. Useful statistics to describe the pulse rate data are the range and mean of the data.

The range of this data set can be expressed as 'between x beats per minute (lowest value) and y beats per minute (highest value).'

The mean is the total of all the values divided by the number of data points. It is an estimate of the true value of the variable being measured.

Q Write down two statements about the pulse rate data.

Q If you are comparing the pulse rates of two different classes in your school, why would it be useful to have both statistics (mean and range)?

Interpreting scale visuals

It is important to note the scale of the picture, and to understand units of measurement.

Q What is the actual size of a starch grain shown in the photograph on the right?

Q Which is longer, 1000 nm or 10 μm?

Q Describe the three ways that magnification is shown in the photographs on this page.

Pollen contains the male gametes of a flowering plant.

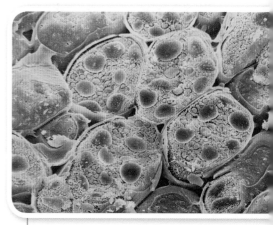

Starch grains in a plant cell store glucose as starch. (Magnification × 200.)

Controlled assessment

In GCSE Biology the controlled assessment counts for 25% of your total grade. Marks are given for a practical investigation.

Your school or college may give you the mark schemes for this.

This will help you understand how to get the most credit for your work.

Practical investigation (25%)

Investigations are carried out by scientists to try and find the answers to scientific questions. The skills you learn from this work will help prepare you to study any science course after GCSE.

To succeed with any investigation you will need to:
- choose a question to explore
- select equipment and use it appropriately and safely
- design ways of making accurate and reliable observations
- relate your investigation to work by other people investigating the same ideas.

Your investigation report will be based on the data you collect from your own experiments. You will also use information from other people's research. This is called secondary data.

You will write a full report of your investigation. Marks will be awarded for the quality of your report. You should:
- make sure your report is laid out clearly in a sensible order
- use diagrams, tables, charts, and graphs to present information
- take care with your spelling, grammar, and punctuation, and use scientific terms where they are appropriate.

Marks will be awarded under five different headings.

Strategy
- Develop a hypothesis to investigate.
- Choose a procedure and equipment that will give you reliable data.
- Carry out a risk assessment to minimise the risks of your investigation.
- Describe your hypothesis and plan using correct scientific language.

Collecting data
- Carry out preliminary work to decide the range.
- Collect data across a wide enough range.
- Collect enough data and check its reliability.
- Control factors that might affect the results.

Analysis
- Present your data to make clear any patterns in the results.
- Use graphs or charts to indicate the spread of your data.
- Use appropriate calculations such as averages and gradients of graphs.

Evaluation
- Describe and explain how you could improve your method.
- Discuss how repeatable your evidence is, accounting for any outliers.

Review
- Comment, with reasons, on your confidence in the secondary data you have collected.
- Compare the results of your investigation to the secondary data.
- Suggest ways to increase the confidence in your conclusions.

Secondary data

Once you have collected the data from your investigation you should look for some secondary data relevant to your hypothesis. This will help you decide how well your data agrees with the findings of other scientists. Your teacher will give you secondary data provided by OCR, but you should look for further sources to help you evaluate the quality of all your data. Other sources of information could include:

- experimental results from other groups in your class or school
- text books
- the Internet.

When will you do this work?

Your school or college will decide when you do your practical investigation. If you do more than one investigation, they will choose the one with the best marks.

Your investigation will be done in class time over a series of lessons.

You may also do some research out of class.

B1 You and your genes

Why study genes?

What makes me the way that I am? Your ancestors probably asked the same question. How are features passed on from parents to children? You may look like your relatives, but you are unique. Only in the last few generations has science been able to answer questions like these.

What you already know

- In sexual reproduction fertilisation happens when a male and female sex cell join together. Information from two parents is mixed to make a new plan for the offspring. The offspring will be similar but not identical to their parents.

- There are variations between members of the same species that are due to environmental as well as inherited causes.

- Clones are individuals with indentical genetic information.

- The science of cloning raises ethical issues.

Find out about

- how genes and your environment make you unique

- how and why people find out about their genes

- how we can use our knowledge of genes

- whether we should allow this.

The Science

Your environment has a huge effect on you, for example, on your appearance, your body, and your health. But these features are also affected by your genes. In this Module you'll find out how. You'll discover the story of inheritance.

Ideas about Science

In the future, science could help you to change your baby's genes before it is born. Cloned embryos could provide cells to cure diseases. But, as new technologies are developed, we must decide how they should be used. These can be questions of ethics – decisions about what is right and wrong.

Find out about

- ✓ **what makes us all different**
- ✓ **what genes are and what genes do**

Plants and animals look a lot like their parents. They have **inherited** information from them. This information is in **genes** and controls how the organisms develop and function.

A lot of information goes into making a human being. So inheritance does a big job pretty well. All people have most features in common. Children look a lot like their parents. If you look at the people around you, the differences between us are very, very small. But we're interested in them because they make us unique.

Most features are affected by both the information you inherit and your environment.

These sisters have some features in common.

Environment makes a difference

Almost all of your features are affected by the information you inherited from your parents. For example, your blood group depends on this information. Some features are the result of only your environment, such as scars and tattoos.

But most of your features are affected by both your genes and your **environment**. For example, For example, your weight depends on inherited information, but if you eat too much, you become heavier

Key words

- ✓ **inherited**
- ✓ **genes**
- ✓ **environment**

Questions

1 Choose two of the students in the photograph on the left. Write down five ways they look different.

2 What two things can affect how you develop?

3 Explain what is meant by inherited information.

Inheritance – the story of life

One important part of this story is where all the information is kept. Living organisms are made up of cells. If you look at cells under a microscope you can see the **nuclei**. Inside each nucleus are long threads called **chromosomes**. Each chromosome contains thousands of genes. It is genes that control how you develop.

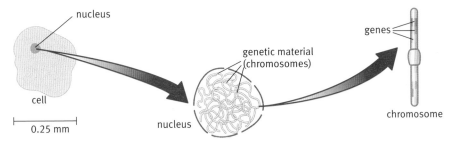

| All the information needed to create a whole human being fits into the nucleus of a cell. The nucleus is just 0.006 mm across!

What are chromosomes made of?

Chromosomes are made of very long molecules of **DNA**. DNA is short for deoxyribonucleic acid. A gene is a section of a DNA molecule.

How do genes control your development?

A fertilised egg cell has the instructions for making every **protein** in a human being. That's what genes are – instructions for making proteins. Each gene is the instruction for making a different protein.

What's so important about proteins?

Proteins are important chemicals for cells. There are many different proteins in the body, and each one does a different job. They may be:
- **structural** proteins – to build the body, eg collagen, the protein found in tendons
- **functional** proteins – to take part in the chemical reactions of the body, for example, enzymes such as amylase.

Genes control which proteins a cell makes. This is how they control what the cell does and how an organism develops.

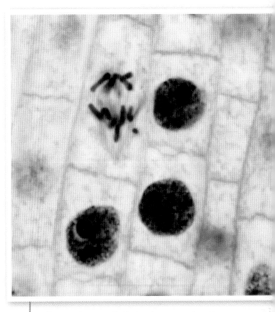

These plant cells have been stained to show up their nuclei. One cell is dividing and the separating chromosomes can be seen.

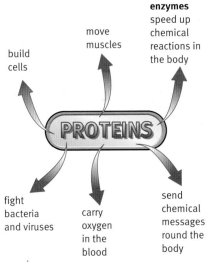

There are about 50 000 types of proteins in the human body.

Questions

4 Write these cell parts in order starting with the smallest: chromosome, gene, cell, nucleus

5 Explain how genes control what a cell does.

6 a List two kinds of job that proteins do in the human body.
  b Name two proteins in the human body and say what they do.

Key words
- ✔ nuclei
- ✔ chromosomes
- ✔ structural
- ✔ DNA
- ✔ protein
- ✔ enzymes
- ✔ functional

Find out about

- ✔ **why identical twins look like each other**
- ✔ **why identical twins do not stay identical**
- ✔ **what a clone is**

When a baby is born, who can say how it will grow and develop? Your genes decide a lot about you. A few characteristics, like having dangly earlobes or dimples, are decided by just one pair of genes. Most of your characteristics, such as your height, your weight, and your eye colour, are decided by several different genes working together. But your genes don't tell the whole story.

Twins and the environment

○ Identical twins have the same genes but they don't look exactly the same.

Identical twins are formed when a fertilised egg starts to divide and splits to form two babies instead of one. Each baby has the same genetic information. This means any differences between them are because of the environment.

Most identical twins are brought up together in the same family so their environment is very similar. But sometimes twins are separated after birth and adopted by different parents. This allows scientists to find out how much their characteristics are because of their genes, and how much they are affected by the different environments they live in.

It is often surprising how alike the separated twins are when they meet. The influence of the genes is very strong. The different environments do mean that some things, like the weights of the twins, are more different than for twins who are brought up together.

Identical twins that are separated at birth are not very common. It would be wrong for scientists to separate babies just to find out how much effect the environment has on how they grow and develop. For this reason scientists often study plants instead.

Questions

1 What are the differences between the ways that having dimples, having green eyes, and being 2 m tall are determined?

2 Why do scientists find studying identical twins so useful?

Cloning

We call any genetically identical organisms **clones**. So identical twins are human clones! Plant clones are quite common. Strawberry plants and spider plants are just two sorts of plant that make identical plant clones at the end of runners. Bulbs, like daffodils, also produce clones.

It is easy for people to clone plants artificially. This can be done by taking cuttings. A piece of the adult plant is cut off. It soon forms new roots and stems to become a small plant genetically identical to the original parent. The new plant is a clone – it has the same genes as the parent plant.

You can also make clones from tiny pieces of plants grown in special jelly, called agar. In this way you can make hundreds of clones from a single plant.

Once you have some clones, you can look at how the environment affects them. If the parent plant grew very tall, that will be partly down to its genes. But what happens if it doesn't get enough nutrients, or it is short of water? Will it still grow tall? When we look at the effect of different factors on the characteristics of cloned plants, it helps us to understand how genes and the environment interact.

Each of these baby spider plants is a clone of its parent plant and of all the other baby plants.

Questions

3 What is a clone?

4 Why are cloned plants so useful to scientists?

5 Describe how you could use cloned plants to show how the environment affects their appearance.

Key word

✓ **clone**

You may make cauliflower clones like these.

What makes you the way you are?

Find out about

- ✔ **how you inherit genes**
- ✔ **Huntington's disease (an inherited illness)**

It can be funny to see that people in a family look like each other. Perhaps you don't like a feature you've inherited – your dad's big ears or your mum's freckles. For some people, family likenesses are very serious.

Craig's story

My grandfather's only 56. He's always been well but now he's a bit off colour. He's been forgetting things – driving my Nan mad. No-one's said anything to me, but they're all worried about him.

Robert's story

I'm so frustrated with myself. I can't sit still in a chair. I'm getting more and more forgetful. Now I've started falling over for no reason at all. The doctor has said it might be **Huntington's disease**. It's an inherited disorder. She said I can have a blood test to find out, but I'm very worried.

Huntington's disease

Huntington's disease is an inherited disorder. You can't catch it. The disease is passed on from parents to their children. The symptoms of Huntington's disease don't appear until middle age. First the person has problems controlling their muscles. This shows up as twitching. Gradually a sufferer becomes forgetful. They find it harder to understand things and to concentrate. People with Huntington's often have mood changes. After a few years they can't control their movements. Sadly, the condition is fatal.

○ Craig and his grandfather, Robert.

Key word

- ✔ **Huntington's disease**

Questions

1 List the symptoms of Huntington's disease.

2 Explain why Huntington's disease is called an inherited disorder.

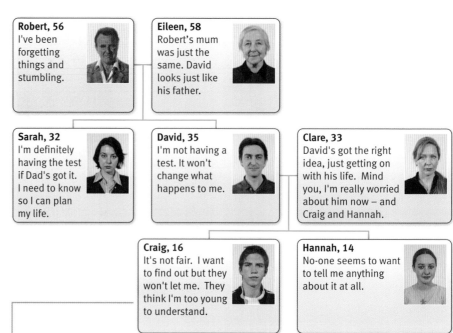

Robert, 56
I've been forgetting things and stumbling.

Eileen, 58
Robert's mum was just the same. David looks just like his father.

Sarah, 32
I'm definitely having the test if Dad's got it. I need to know so I can plan my life.

David, 35
I'm not having a test. It won't change what happens to me.

Clare, 33
David's got the right idea, just getting on with his life. Mind you, I'm really worried about him now – and Craig and Hannah.

Craig, 16
It's not fair. I want to find out but they won't let me. They think I'm too young to understand.

Hannah, 14
No-one seems to want to tell me anything about it at all.

○ Craig's family tree.

How do you inherit your genes?

Sometimes people in the same family look a lot alike. In other families brothers and sisters look very different. They may also look different from their parents. The key to this mystery lies in our genes.

Parents pass on genes in their **sex cells**. In animals these are sperm and egg cells (ova). Sex cells have copies of half the parent's chromosomes. When a sperm cell fertilises an egg cell, the fertilised egg cell gets a full set of chromosomes. It is called an **embryo**.

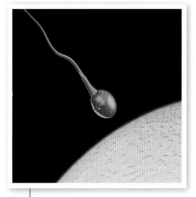

The fertilised egg cell will have genes from both parents. (Mag: × 2000 approx.)

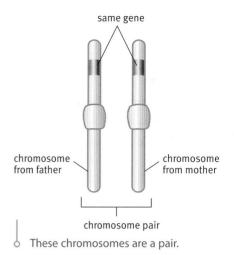

same gene

chromosome from father

chromosome from mother

chromosome pair

These chromosomes are a pair.

How many chromosomes does each cell have?

Chromosomes come in pairs. Every human body cell has 23 pairs of chromosomes. The chromosomes in most pairs are the same size and shape. They carry the same genes in the same place. This means that your genes also come in pairs.

Sex cells have single chromosomes

Sex cells are made with copies of half the parent's chromosomes, one from each pair. This makes sure that the fertilised egg cell has the right number of chromosomes – 23 pairs.

One chromosome from each pair came from the egg cell. The other came from the sperm cell. Each chromosome carries thousands of genes. Each chromosome in a pair carries the same genes along its length.

So the fertilised egg cell has a mixture of the parents' genes. Half of the new baby's genes are from the mother. Half are from the father. This is why children resemble both their parents.

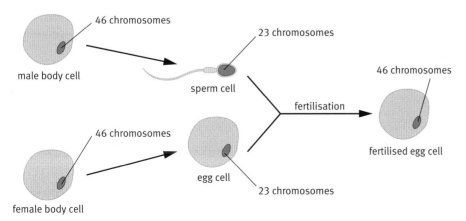

46 chromosomes

male body cell

23 chromosomes

sperm cell

46 chromosomes

female body cell

egg cell

23 chromosomes

fertilisation

46 chromosomes

fertilised egg cell

The cells in this diagram are not drawn to scale. A human egg cell is 0.1 mm across. This is 20 times larger than a human sperm cell.

Key words
- ✔ **sex cells**
- ✔ **embryo**

Questions

3 a Draw a diagram to show a sperm cell, an egg cell, and the fertilised egg cell they make.

  b Explain why the fertilised egg cell has pairs of chromosomes.

4 Explain why children may look a bit like each of their parents.

5 Two sisters with the same parents won't look exactly alike. Explain why this is.

Find out about

- ✔ **what decides if you are male or female**
- ✔ **how a Y chromosome makes a baby male**

Ever wondered what it would be like to be the opposite sex? Well, if you are male there was a time when you were female – just for a short while. Male and female embryos are very alike until they are about six weeks old.

What decides an embryo's sex?

A fertilised human egg cell has 23 pairs of chromosomes. Pair 23 are the sex chromosomes. Males have an X chromosome and a Y chromosome – **XY**. Females have two X chromosomes – **XX**.

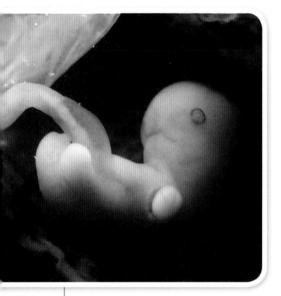

This embryo is six weeks old.

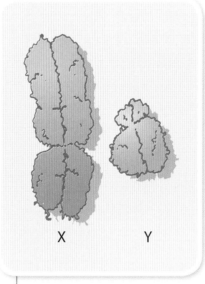

X   Y

Women have two X chromosomes. Men have an X and a Y.

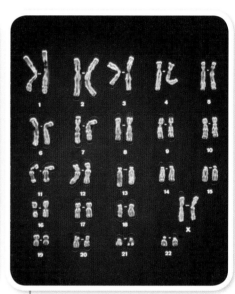

These chromosomes are from the nucleus of a woman's body cell. They have been lined up to show the pairs.

Key words

- ✔ **XY chromosomes**
- ✔ **XX chromosomes**

What's the chance of being male or female?

A parent's chromosomes are in pairs. When sex cells are made they only get one chromosome from each pair. So half a man's sperm cells get an X chromosome and half get a Y chromosome. All a woman's egg cells get an X chromosome.

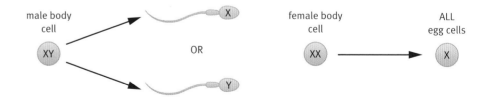

When a sperm cell fertilises an egg cell the chances are 50% that it will be an X or a Y sperm. This means that there is a 50% chance that the baby will be a boy or a girl.

How does the Y chromosome make a baby male?

A male embryo's testes develop when it is about six weeks old. This is caused by a gene on the Y chromosome – the SRY gene. SRY stands for 'sex-determining region of the Y chromosome'.

Testes produce male sex hormones called androgens. Androgens make the embryo develop into a male. If there is no male sex hormone present, the sex organs develop into the ovaries, clitoris, and vagina of a female.

SRY gene

The Y chromosome

What are hormones?

Hormones are chemicals that control many processes in the cells. Tiny amounts of hormones are made by different parts of the body. You can read more about hormones in Module B2: *Keeping Healthy*.

Jan's story

At eighteen Jan was studying at college. She was very happy, and was going out with a college football player. She thought her periods hadn't started because she did a lot of sport.

Then in a science class Jan looked at the chromosomes in her cheek cells. She discovered that she had male sex chromosomes – XY.

Sometimes a person has X and Y chromosomes but looks female. This is because their body makes androgens but the cells take no notice of it. About 1 in 20 000 people have this condition. They have small internal testes and a short vagina. They can't have children.

Jan had no idea she had this condition. She found it very difficult to come to terms with. But she has now told her boyfriend and they have stayed together.

Jan on holiday, aged eighteen.

Questions

1 Which sex chromosome(s) would be in the nucleus of:
   a a man's body cell?
   b an egg cell?
   c a woman's body cell?
   d a sperm cell?

2 Explain why an embryo needs the Y chromosome to become a boy.

3 Imagine you are Jan or her boyfriend. How would you have felt about her condition?

Find out about

- ✓ how pairs of genes control some features
- ✓ cystic fibrosis (an inherited illness)
- ✓ testing a baby's genes before they are born

Key words

- ✓ alleles
- ✓ phenotype
- ✓ dominant
- ✓ genotype
- ✓ recessive

This diagram shows one pair of chromosomes. The gene controlling dimples is coloured in.

Will this baby be tall and have red hair? Will she have a talent for music, sport – even science?! Most of these features will be affected by both her environment and her genes. A few features are controlled by just one gene. We can understand these more easily.

This baby has inherited a unique mix of genetic information.

Genes come in different versions

Both chromosomes in a pair carry genes that control the same features. Each pair of genes is in the same place on the chromosomes.

But genes in a pair may not be exactly the same. They may be slightly different versions. You can think about it like football strips – often a team's home and away strips are both based on the same pattern, but they're slightly different. Different versions of the same genes are called **alleles**.

Dominant alleles – they're in charge

The gene that controls dimples has two alleles. The D allele gives you dimples. The d allele won't cause dimples. The alleles you inherit is called your **genotype**.

Your **phenotype** is what you look like, eg dimples or no dimples, your characteristics.

The D allele is **dominant**. You only need one copy of a dominant allele to have its feature. The d allele is **recessive**. You must have two copies of a recessive allele to have its feature – in this case no dimples.

Which alleles can a person inherit?

Sex cells get one chromosome from each pair the parent has. So they only have one allele from each pair. If a parent has two D or two d alleles, they can only pass on a D or a d allele to their children.

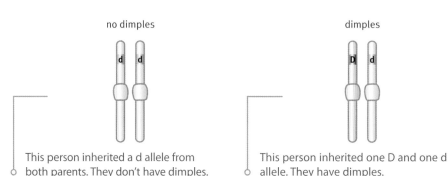

dimples

This person inherited a D allele from both parents. They have dimples.

no dimples

This person inherited a d allele from both parents. They don't have dimples.

dimples

This person inherited one D and one d allele. They have dimples.

But a parent could have one D and one d allele. Then half of their sex cells will get the D allele and half will get the d allele.

The human lottery

We cannot predict which egg and sperm cells will meet at fertilisation. This genetic diagram (Punnett square) shows all the possibilities for one couple.

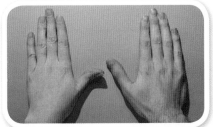

The allele that gives you straight thumbs is dominant (T). The allele for curved thumbs is recessive (t).

The allele that gives you hair on the middle of your fingers is dominant (R). The allele for no hair is recessive (r).

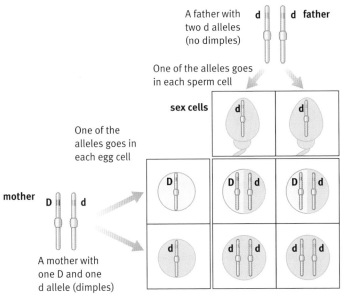

A father with two d alleles (no dimples)

d   d   **father**

One of the alleles goes in each sperm cell

sex cells

One of the alleles goes in each egg cell

mother   D   d

A mother with one D and one d allele (dimples)

children There is a 50% chance of a child having dimples

Why don't brothers and sisters look the same?

Human beings have about 23 000 genes. Each gene has different alleles. If both of the alleles you inherit are the same you are **homozygous** for that characteristic. If you inherit different alleles you are **heterozygous**.

Brothers and sisters are different because they each get a different mixture of alleles from their parents. Except for identical twins, each one of us has a unique set of genes.

What about the family?

Huntington's disease is a single-gene disorder caused by a dominant allele. You only need to inherit the allele from one parent to have the condition. In Craig's family, whom we met on page 20, Craig's grandfather, Robert, has Huntington's disease. So their dad, David, may have inherited this faulty allele. At the moment he has decided not to have the test to find out.

Key words
- **homozygous**
- **heterozygous**

Questions

1 Write down what is meant by the word allele.

2 Explain how you inherit two alleles for each gene.

3 Explain the difference between a dominant and a recessive allele.

4 What are the possible pairs of alleles a person could have for:
   a dimples?
   b straight thumbs?
   c no hair on the second part of their ring finger?

5 Use a diagram to explain why a couple who have dimples could have a child with no dimples.

6 Use a diagram to work out the chance that David has inherited the Huntington's disease allele.

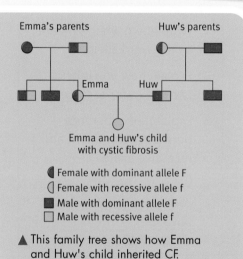

Emma's parents   Huw's parents

Emma   Huw

Emma and Huw's child
with cystic fibrosis

◖ Female with dominant allele F
◖ Female with recessive allele f
■ Male with dominant allele F
□ Male with recessive allele f

▲ This family tree shows how Emma and Huw's child inherited CF.

Dear Doctor

We've had a huge postbag in response to last month's letter from Emma. So this month we're looking in depth at cystic fibrosis, a disease that one in twenty-five of us carries in the UK ...

What is cystic fibrosis?

You can't catch cystic fibrosis. It is a genetic disorder. This means it is passed on from parents to their children. The disease causes big problems for breathing and digestion. Cells that make mucus in the body don't work properly. The mucus is much thicker than it should be, so it blocks up the lungs. It also blocks tubes that take enzymes from the pancreas to the gut. People with cystic fibrosis get breathless. They also get lots of chest infections. The shortage of enzymes in their gut means that their food isn't digested properly. So the person can be short of nutrients.

How do you get cystic fibrosis?

Most people with cystic fibrosis (CF) can't have children. The thick mucus affects their reproductive systems. So babies with CF are usually born to healthy parents. At first glance this seems very strange – how can a parent pass a disease on to their children when they don't have it themselves?

The answer lies with one of the thousands of genes responsible for producing a human being. One of these instructs cells to make mucus. But sometimes there are errors in the DNA, so that it does not do its job.

A person who has one dominant, normal-functioning allele (F) and one recessive, faulty allele (f) will not have CF. They can still make normal mucus. But they are a carrier of the faulty allele. When parents who are carriers make sex cells, half will contain the normal allele – and half will contain the faulty allele. When two sex cells carrying the faulty allele meet at fertilisation, the baby will have CF. One in every 25 people in the UK carries the CF allele.

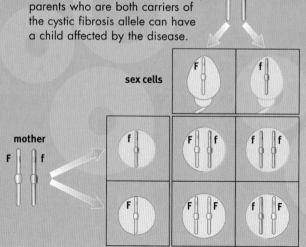

This diagram shows how healthy parents who are both carriers of the cystic fibrosis allele can have a child affected by the disease.

F f father

sex cells

mother
F f

children There is a 25% chance that a child from the carrier parents will have cystic fibrosis.

Key words

- ✓ cystic fibrosis
- ✓ termination
- ✓ carrier

Can cystic fibrosis be cured?

Not yet. But treatments are getting better, and life expectancy is increasing all the time. Physiotherapy helps to clear mucus from the lungs. Sufferers take tablets with the missing gut enzymes in. Antibiotics are used to treat chest infections. And an enzyme spray can be used to thin the mucus in the lungs, so it is easier to get rid of. New techniques may offer hope for a cure in the future.

What are the options?

If a couple know they are at risk of having a child with cystic fibrosis they can have tests to see if their child has the disease. During pregnancy cells from the developing fetus can be collected, and the genes examined. If the fetus has two faulty alleles for cystic fibrosis, the child will have the disease. The parents may choose to end the pregnancy. This is done with a medical operation called a termination (an abortion).

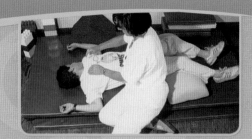

▲ Tom has cystic fibrosis. He has physiotherapy every day to clear thick mucus from his lungs.

How do doctors get cells from the fetus?

The fetal cells can be collected two ways:
- an amniocentesis test
- a chorionic villus test.

The diagrams show how each of these tests is carried out.

▼ Amniocentesis test.

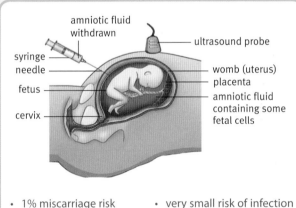

amniotic fluid withdrawn
ultrasound probe
syringe needle
womb (uterus)
placenta
fetus
amniotic fluid containing some fetal cells
cervix

- 1% miscarriage risk
- results at 15–18 weeks
- very small risk of infection
- results not 100% reliable

▼ Chorionic villus test.

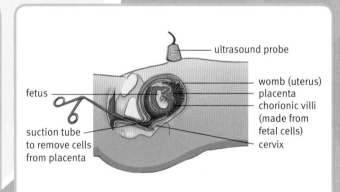

ultrasound probe
fetus
womb (uterus)
placenta
chorionic villi (made from fetal cells)
suction tube to remove cells from placenta
cervix

- 2% miscarriage risk
- results at 10–12 weeks
- almost no risk of infection
- results not 100% reliable

Questions

7 The magazine doctor is sure that nothing Emma did during her pregnancy caused her baby to have cystic fibrosis. How can she be so sure?

8 People with cystic fibrosis make thick, sticky mucus. Describe the health problems that this may cause.

9 Explain what it means when someone is a 'carrier' of cystic fibrosis.

10 Two carriers of cystic fibrosis plan to have children. Draw a diagram to work out the chance that they will have:
a a child with cystic fibrosis
b a child who is a carrier of cystic fibrosis
c a child who has no cystic fibrosis alleles.

Find out about

- **how people make ethical decisions**
- **how genetic information could be used**

'We had an amniocentesis test for each of my pregnancies,' says Elaine. 'Sadly we felt we had to terminate the first one, because the fetus had CF. We are lucky enough now to have two healthy children – and we know we don't have to watch them suffer.'

Key words

- **ethics**
- **genetic test**

Elaine's nephew has cystic fibrosis. When they found out, Elaine and her husband Peter became worried about any children they might have. They both had a **genetic test**. The tests showed that they were both carriers for cystic fibrosis. Elaine and Peter decided to have an amniocentesis test when Elaine was pregnant.

Elaine and Peter made a very hard decision when they decided to terminate their first pregnancy. When a person has to make a decision about what is the right or wrong way to behave, they are thinking about **ethics**. Deciding whether to have a termination is an example of an ethical question.

Ethics – right and wrong

For some ethical questions, the right answer is very clear. For example, should you feed and care for your pet? Of course. But in some situations, like Elaine and Peter's, people may not agree on one right answer. People think about ethical questions in different ways.

For example, Elaine and Peter felt that they had weighed up the consequences of either choice. They thought about how each choice – continuing with the pregnancy or having a termination – would affect all the people involved. They had to make a judgement about the difficulties their unborn child would face with cystic fibrosis.

In order to consider all the consequences they also had to think about the effects that an ill child would have on their lives, and on the lives of any other children they might have. Some people feel that they could not cope with the extra responsibility of caring for a child with a serious genetic disorder.

Different choices

Not everyone weighing up the consequences of each choice would have come to the same decision as this couple did.

Jo has a serious genetic disorder. Her parents believe that termination is wrong. They decided not to have more children, rather than use information from an amniocentesis test.

Some people feel that any illness would have a devastating effect on a person's quality of life. But some people lead very happy, full lives in spite of very serious disabilities.

Elaine and Peter made their ethical decision only by thinking about the consequences that each choice would have. This is just one way of dealing with ethical questions.

When you believe that an action is wrong

For some people having a termination would be completely wrong in itself. They believe that an unborn child has the right to life, and should be protected from harm in the same way as people are protected after they are born. Other people believe that terminating a pregnancy is unnatural, and that we should not interfere. People may hold either of these viewpoints because of their own personal beliefs, or because of their religious beliefs.

A couple in Elaine and Peter's position who felt that termination was wrong might decide not to have children at all. This would mean that they could not pass on the faulty allele. Or they could decide to have children, and to care for any child that did inherit the disease.

What are the ethical arguments for a decision?

The right decision is the one which leads to the best outcome for the most people.

Some actions are wrong and should never be done.

It's wrong to have a termination. We'll look after our baby whatever.

Is it fair for us to have a baby knowing they'll be ill?

Questions

1   Explain what is meant by 'an ethical question'.

2   Describe three different points of view that a couple in Elaine and Peter's position might take.

3   What is your viewpoint on genetic testing of a fetus for a serious illness? Explain why you think this.

Find out about

- ✓ **what a genetic test is**
- ✓ **what genetic screening is**

How reliable are genetic tests?

Genetic testing is used to look for alleles that cause genetic disorders. People like Elaine and Peter use this information to make decisions about whether to have children. Genetic tests can also be used to make a decision about whether a pregnancy should be continued or not.

So, it is important to realise that the tests are not completely reliable. In a few cases only it will not detect CF. This is called a **false negative**. The test only looks for common DNA errors in the faulty CF gene. **False positive** tests are not common, but they may happen due to technical failure of the test.

Why do people have genetic tests?

Usually people only have a genetic test because they know that a genetic disorder runs in their family; they want to know if they are carriers. Most parents who have a child with cystic fibrosis did not know that they were carriers. So, they would not have had a genetic test during pregnancy.

Every newborn baby in the UK is now screened for cystic fibrosis. They have a blood test – this is a biochemical test, not the more expensive genetic test. If the biochemical test is positive for CF the baby will be genetically tested to confirm the diagnosis. Treatment can start before the lungs are too badly damaged. Gentetic testing the whole population or large groups for a genetic disease is called **genetic screening**.

Who decides about genetic screening?

The decision about whether to use genetic screening is taken by governments and local NHS trusts. People in the NHS have to think about different things when they decide if genetic screening should be used:

- What are the costs of testing everyone?
- What are the benefits of testing everyone?
- Is it better to spend the money on other things, for example, hip replacement operations, treating people with cancer, and treating people who already have cystic fibrosis?

Can we? Should we?

Four of Rabbi Joseph Ekstein's children died from a severe genetic condition called Tay-Sachs disease. In the general population about one baby in every 300 000 has Tay-Sachs, but in the 1980s one baby in every 3600 born to European Jewish families was affected and died.

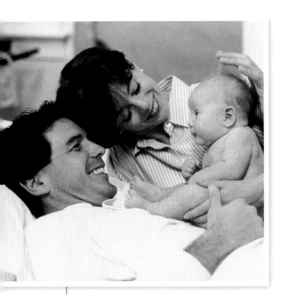

This couple are both carriers of cystic fibrosis. They had an amniocentesis test during their pregnancy. The results showed that the baby did not have CF. When their daughter was born she was completely healthy.

Key words

- ✓ **false negative**
- ✓ **false positive**
- ✓ **genetic screening**

In 1983 Rabbi Ekstein set up a genetic screening programme. Couples planning to marry were genetically tested. If they both carried the recessive allele they were advised either to not marry or to have prenatal screening and terminate affected pregnancies. As a result of this genetic screening, Tay-Sachs has almost disappeared from Jewish communities worldwide.

Testing your genes

If you could have more information about your genes, would you want it? You can buy a DNA testing kit that tells you if you are carrying faulty alleles for over 100 rare genetic mutations including cystic fibrosis and Tay-Sachs. Some scientists hope tests like these will help to prevent many genetic diseases. Other scientists think that the risk of being affected by these rare genetic diseases is so low that screening is not worthwhile. It costs money and may cause people to worry unnecessarily.

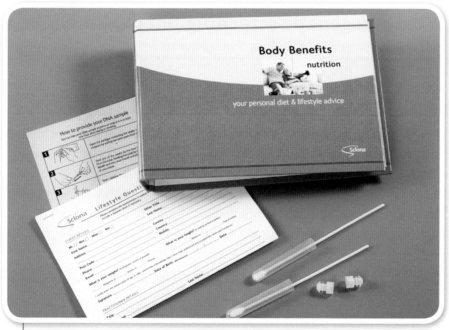

DNA testing can already be done at homes with simple kits like this. How much do you want to know about your DNA?

In the future

Scientists can already work out the complete DNA sequence (the genome) of anyone who has enough money to pay thousands of pounds. In five years' time it may be so cheap that everyone will be able to have it done. The genome of every newborn baby may be worked out. How might we use this information?

Questions

1 What are 'false negative' and 'false positive' results?

2 Why is it important for people to know about false results?

3 Explain what is meant by the term 'genetic screening'.

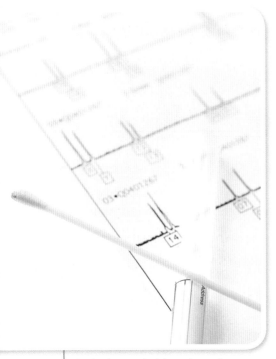

Carolyn had a dangerous reaction to drugs. Genetic testing may help to avoid this.

Finding the right medicine

In 2009 Carolyn Major started to take the medicine she hoped would help cure her cancer. Instead, four days later, she was in an intensive care ward with her heart struggling to keep going. She is one of a small group of people whose bodies react very badly to the anti-cancer drug. Carolyn was lucky. She recovered with no permanent damage done to her heart – and so far the cancer hasn't returned either.

In future the genetic testing of adults may make problems like this a thing of the past. Doctors think that we can use genetic testing to match medicines to patients. Some people produce enzymes that break down drugs very quickly, so they need higher doses of a medicine than most of us. Other people don't produce enzymes that break down certain drugs in their body, so they can be poisoned by medicine that is meant to help them. In future, genetic testing may show if people have the genes for these enzymes so we can all be given the drugs that work best for our bodies' genotype.

What can genetic screening tell us?

Full genetic screening will show if you have inherited genes that increase your risk of a particular problem such as heart disease or different types of cancer. But remember that most of these diseases are affected by both your genes and your lifestyle.

This information could be very helpful. If you found out that you are genetically at a higher than average risk of developing heart disease, for example, you might decide to change your lifestyle. If you don't smoke, eat a healthy diet, and always take lots of exercise, you would lower your environmental risk of heart disease. This would help to balance your increased genetic risk. And if you know that your genotype means you have an increased risk of getting a particular type of cancer, you can be screened regularly to catch the disease as early as possible if it develops.

Questions

4 How can the genetic testing of adults prevent the birth of babies affected with genetic diseases?

5 What do you think are the advantages and disadvantages of testing adults?

6 How might genetic testing make medicines more effective and what problems might it cause?

In the future, decisions about drug treatments may be based on genetic test results.

Is it right to use genetic screening?

It is easy to see why people may want genetic screening:

* when two people decided to have children, they would know if their children were at risk of inheriting the disorder
* when a disease runs in someone's family, they would know if they were going to develop it.

At first glance, genetic screening may seem like the best course of action for everyone. But the best decision for the majority is not always the right decision. There are ethical questions to consider about genetic screening, including:

* Who should know the test results?
* What effect could the test result have on people's future decisions?
* Should people be made to have screening, or should they be able to opt out? Is it right to interfere?

About 1 in 25 people in the UK carry the allele for cystic fibrosis. Some people think that having this information is useful, but there are also good reasons why not everyone agrees. A decision may benefit many people. But it may not be the right decision if it causes a great amount of harm to a few people.

People have different ideas about whether genetic screening for cystic fibrosis would be a good thing.

Questions

7 Give two arguments for and two against genetic screening for cystic fibrosis.

8 Which argument do you agree with? Explain why.

Who should know about your genes?

If you find out that your genotype suggests you have a higher-than-average risk of developing heart disease, who should know about it? Your partner? Your family? Your boss? Your insurance company?

Many people think that only you and your doctor should know information about your genes. They are worried that it could affect a person's job prospects and chances of getting life insurance.

How does life insurance work?

People with life insurance pay a regular sum of money to an insurance company. This is called a premium. In return, when they die, the insurance company pays out an agreed sum of money. People buy life insurance so that there will be money to support their families when they die.

People use information like this to decide the premium each person should pay for insurance. The higher the risk, the bigger the premium.

Condition	Percentage of deaths from each disease caused by smoking in 2007	
	Men	Women
Cancers		
Lung	88	75
Throat and mouth	76	56
Oesophagus	70	63
Heart and circulation diseases	11	15

Should insurance companies know about your genes?

Insurance companies assess what a person's risk is of dying earlier than average. If they believe that the risk is high, they may choose to charge higher premiums than average. Some people think that insurance companies might use the results of genetic tests in the wrong way. Individual people might do the same. But having a gene that means you have a higher risk does not mean that you will definitely die early. Here are some of the arguments:

- Insurers may not insure people if a test shows that they are more likely to get a particular disease. Or they may charge a very high premium.
- Insurers may say that everyone must have genetic tests for many diseases before they can be insured.
- People may not tell insurance companies if they know they have a genetic disorder.
- People may refuse to have a genetic test because they fear that they will not be able to get insurance. They may miss out on medical treatment that could keep them healthy.

In the USA, the 2008 Genetic Information Nondiscrimination Act means insurance companies can't stop someone getting health insurance, or even charge them higher premiums, because of the results of a genetic test. It also means that employers can't use genetic information when they are deciding who should get a job or who should be fired.

In the UK insurance companies agreed not to collect and use genetic information about people until the end of 2014. There are also some concerns about whether the results of genetic testing might affect employment. Politicians are considering bringing in laws to control how information about people's genes can be used. This debate has been going on for 10 years now and no decisions have been made yet!

What about the police?

Scientists already use information about people's DNA to help them solve crimes. They produce DNA profiles from cells left at a crime scene. There is usually only a 1 in 50 million chance of two people having the same DNA profile – unless they are identical twins.

Some people think that there should be a national DNA database recording everyone living in the country. If all babies were to have their genome sequenced at birth, this will be easily available in the future. At the moment only DNA from people who have been arrested or convicted of committing a crime can be stored. Human rights campaigners are against a database for everyone's DNA. They think it is an invasion of privacy and puts innocent people on the same level as criminals.

DNA evidence has been used to prove many people guilty – and others innocent.

> ## Questions
>
> 9 If genetic testing shows you have a higher-than-average risk of heart disease, do you think this should affect your application:
>   a to be an airline pilot?
>   b to take out life insurance?
>
> 10 Write down arguments for and against the government setting up a DNA database.
>
> 11 Do you agree or disagree with the genetic testing of newborn babies and the setting up of a DNA database? Explain your reasons.

Sally takes a 'fertility drug' so that she releases several eggs. Fertility drugs contain hormones.

↓

In a small operation, the doctor collects the eggs.

↓

Bob's sperm fertilise the eggs in a Petri dish. This is in vitro fertilisation.

↓

When the embryos reach the eight-cell stage, one cell is removed from each.

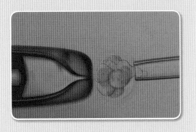

↓

The cells are tested for the Huntington's allele. This is called pre-implantation genetic diagnosis (PGD).

↓

Only embryos without the Huntington's allele are implanted in Sally's uterus.

Many people do not agree with termination. If they are at risk of having a child with a genetic disease, they may have decided not to have children. Now doctors can offer them another treatment. It uses in vitro fertilisation (IVF). In this treatment the mother's egg cells are fertilised outside her body. This treatment is used to help couples who cannot conceive a child naturally. It can also be used to help couples whose children are at risk from a serious genetic disorder.

Pre-implantation genetic diagnosis

Bob and Sally want children, but Bob has the allele for Huntington's disease. Sally has become pregnant twice. Tests showed that both the fetuses had the Huntington's allele and the pregnancies were terminated. Bob and Sally were keen to have a child, so their doctor suggested that they should use **pre-implantation genetic diagnosis** (**PGD**). Sally's treatment is explained in the flow chart. The first use of PGD to choose embryos was in the UK in 1989. At the moment, PGD is only allowed for families with particular inherited conditions.

New technology – new decisions

New technologies like PGD often give us new decisions to make. In the UK, Parliament makes laws to control research and technologies to do with genes. Scientists are not free to do whatever research they may wish to do. From time to time Parliament has to update the law.

But Parliament can't make decisions case by case. So the Government has set up groups of people to decide which cases are within the law on reproduction. One of these groups **regulates** UK fertility clinics and all UK research involving embryos..

The regulatory body interprets the laws we already have about genetic technologies. It also takes into account public opinion, as well as practical and ethical considerations. One of its jobs is to decide when PGD can be used.

If embryos are tested before implantation, no babies need to be born with the disease. The embryos with faulty genes are destroyed. However, some people are concerned about this.

Questions

1 Explain why everyone who has PGD has to use IVF treatment in order to conceive.

2 Make a list of viewpoints for and against embryo selection.
   a Which viewpoint do you agree with? Give your reasons.
   b Which viewpoint do you disagree with? Give your reasons.

Cloning: a natural process

Many living things only need one individual to reproduce. This is called **asexual reproduction**. Single-celled organisms like the bacteria in the picture use asexual reproduction.

The bacterium divides to form two new cells. The two cells' genes are identical to each other's. We call genetically identical organisms **clones**. The only variation between them will be caused by differences in their environment.

Asexual reproduction

Larger plants and animals have different types of cells for different jobs. As an embryo grows, cells become **specialised**. Some examples are blood cells, muscle cells, and nerve cells.

Plants keep some unspecialised cells all their lives. These cells can become anything that the plant may need. For example, they can make new stems and leaves if the plant is cut down. These cells can also grow whole new plants. So they can be used for asexual reproduction.

Some simple animals, like the *Hydra* in the picture opposite, also use asexual reproduction. But cloning is very uncommon in animals.

Sexual reproduction

Most animals and plants use **sexual reproduction**. The new offspring have two parents so they are not clones. But clones are sometimes produced – we call them identical twins.

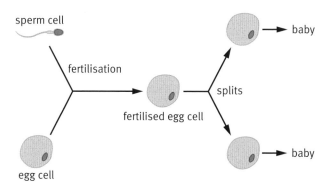

Identical twins have the same genes. But their genes came from both parents. So they are clones of each other, but not of either parent.

Cloning Dolly

Scientists can also clone adult animals artificially but this is much more difficult. Dolly the sheep was the first cloned sheep to be born.
- The nucleus is taken from an unfertilised sheep egg cell.
- The nucleus is taken out of a body cell from a different sheep.

Find out about

- ✔ asexual reproduction
- ✔ cloning and stem cells

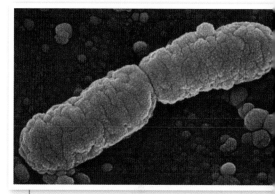

A bacterium cell grows and then splits into two new cells.
(Mag: × 7500 approx.)

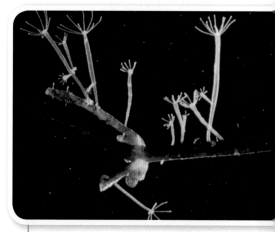

Hydra.

Key words

- ✔ **pre-implantation genetic diagnosis (PGD)**
- ✔ **regulates**
- ✔ **asexual reproduction**
- ✔ **specialised**
- ✔ **sexual reproduction**
- ✔ **clones**

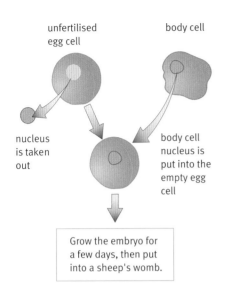

unfertilised egg cell

body cell

nucleus is taken out

body cell nucleus is put into the empty egg cell

Grow the embryo for a few days, then put into a sheep's womb.

Cells from eight-cell embryos like this one can develop into any type of body cell. They start to become specialised when the embryo is five days old. (Mag: × 500 approx.)

- This body cell nucleus is put into the empty egg cell.
- The cell grows to produce a new animal. Its genes will be the same as those of the animal that donated the nucleus. So it will be a clone of that animal.

Dolly died in 2003, aged 6. The average lifespan for a sheep is 12–14 years. Perhaps Dolly's illness had nothing to do with her being cloned. She might have died early anyway. One case is not enough evidence to decide either way.

Many other cloned animals have suffered unusual illnesses. So scientists think that more research needs to be done before cloned mammals will grow into healthy adults.

Cloning humans

In the future it may be possible to clone humans. But most scientists don't want to clone adult human beings. However, some scientists do want to clone human cells. They think that some cloned cells could be used to treat diseases. The useful cells are called **stem cells**.

What are stem cells?

Stem cells are **unspecialised** cells. All the cells in an early embryo are stem cells. These embryonic stem cells can grow into any type of cell in the human body. Stem cells can be taken from embryos that are a few days old. Researchers use human embryos that are left over from fertility treatment.

Adults also have stem cells in many tissues, for example, in their bone marrow, brain, and heart. These unspecialised cells can develop into many, but not all, types of cells. Bone marrow stem cells are already used in transplants to treat patients with leukaemia.

Scientists want to grow stem cells to make new cells to treat patients with some diseases. For example, new brain cells could be made for patients with Parkinson's disease. But these new cells would need to have the same genes as the person getting them as a treatment. When someone else's cells are used in a transplant they are rejected.

What's cloning got to do with this?

Cloning could be used to produce an embryo with the same genes as the patient. Stem cells from this embryo would have the same genes as the patient. So cells produced from the embryo would not be rejected by the patient's body. Cloning a patient's own adult stem cells would produce cells that could be used to treat their illness. The cells would not be rejected because they are the same as the patient's own cells.

Doctors are exploring these technologies. Success could benefit millions of people if it is made to work.

Should human embryo cloning be allowed?

"With some things there's no argument. Murder is just wrong – in the same way that lying and stealing are wrong. Killing an embryo at any age is as wrong as killing a child or an adult."

"Research on embryos is legal up to 14 days. If something is 'legal' it can't be wrong."

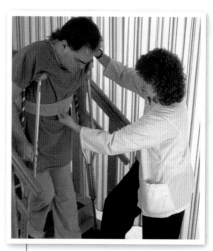

James has Parkinson's disease. His brain cells do not communicate with each other properly. He cannot control his movements.

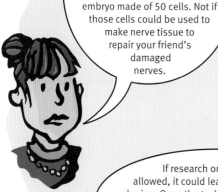

"Whether it's right or not depends on how much good it does versus how much harm. If your best friend was paralysed in an accident, you wouldn't think it was wrong to sacrifice a five-day-old embryo made of 50 cells. Not if those cells could be used to make nerve tissue to repair your friend's damaged nerves."

"An embryo is human so it has human rights. Its age doesn't make any difference. You can't experiment on a child or an adult."

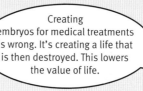

"Creating embryos for medical treatments is wrong. It's creating a life that is then destroyed. This lowers the value of life."

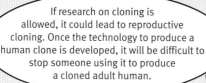

"If research on cloning is allowed, it could lead to reproductive cloning. Once the technology to produce a human clone is developed, it will be difficult to stop someone using it to produce a cloned adult human."

Questions

1 How are stem cells different from other cells?

2 Explain why scientists think stem cells would be useful in treating Parkinson's disease.

3 For each of these cells, say whether or not your body would reject it:
   a bone marrow from your identical twin
   b your own skin cells
   c a cloned embryo stem cell.

4 For embryo cloning to make stem cells:
   a describe one viewpoint in favour
   b describe two different viewpoints against.

5 People often make speculations when they are arguing for or against something. This is something they think will happen, but may not have evidence for. Write down a viewpoint on human cloning that is a speculation.

Science Explanations

In this module, you will learn about inheritance, that genes are the units of inheritance, the relationship between genes and the environment, and that sexual reproduction is a source of variation.

- that genes are sections of DNA and form part of chromosomes that are found in cells' nuclei and instruct cells to make proteins
- how single genes can determine some human characteristics, such as dimples; several genes working together determine many characteristics, such as eye colour; features such as scars are determined by the environment, and other characteristics, such as weight, are determined by both genes and the environment
- that a pair of chromosomes carries the same gene in the same place and alleles are different versions of a gene
- about dominant and recessive alleles and why the recessive gene in a heterozygous pair will not show its characteristics
- that a person's genotype is the genes they have, whilst their phenotype describes their characteristics
- how sex cells contain one chromosome from each pair and during sexual reproduction genes from both parents come together
- why offspring are similar to both parents but are not the same because they inherit a different combination of genes
- how a gene on the Y chromosome determines maleness
- that genetic diagrams are used to show inheritance
- how the symptoms of Huntington's disease appear later in life and include clumsiness, tremors, memory loss, and mood changes and that it is caused by a faulty dominant allele
- how a faulty recessive allele causes cystic fibrosis, a condition where cells produce thick mucus that causes chest infections and difficulties breathing and digesting food
- how genetic testing is used to screen adults, children, and embryos for faulty alleles
- how the information from genetic testing is used to make decisions and why this has implications
- that clones are organisms with identical genes
- about unspecialised cells, called stem cells, that can develop into other types of cell and how they can be used to treat some illnesses.

structural proteins

genes

proteins as enzymes

Huntington's

cystic fibrosis

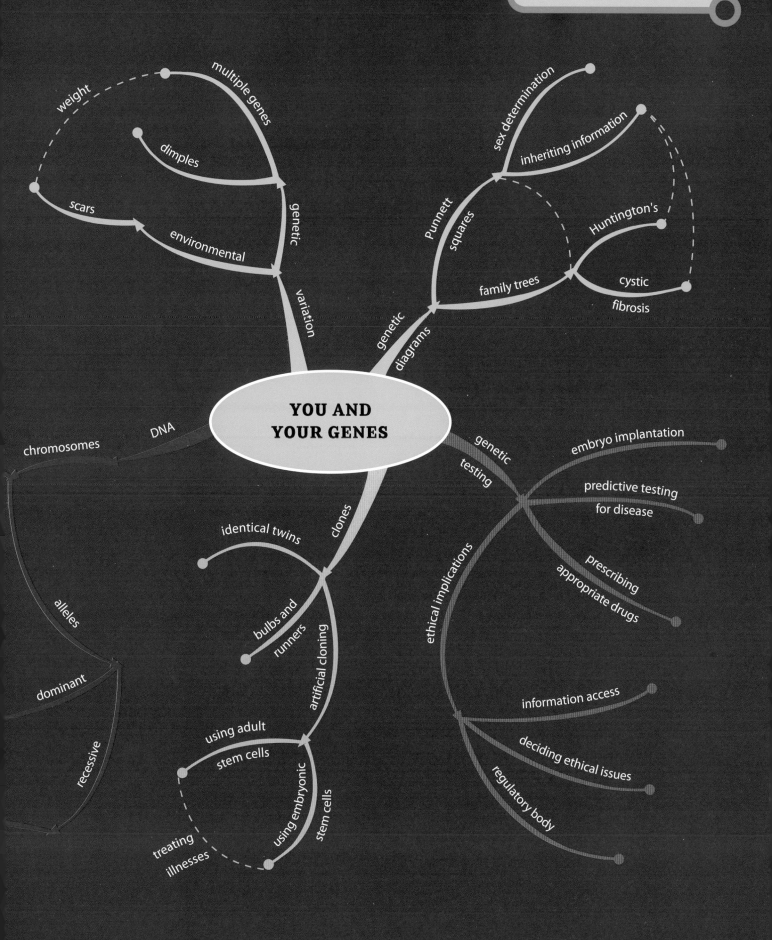

YOU AND YOUR GENES

weight
multiple genes
dimples
scars
environmental
genetic
variation

sex determination
inheriting information
Punnett squares
Huntington's
family trees
cystic fibrosis
genetic diagrams

DNA
chromosomes
alleles
dominant
recessive

genetic testing
embryo implantation
predictive testing for disease
prescribing appropriate drugs
ethical implications
information access
deciding ethical issues
regulatory body

clones
identical twins
bulbs and runners
artificial cloning
using adult stem cells
using embryonic stem cells
treating illnesses

Ideas about Science

The application of science and technology has many implications for society. Ethical issues are often raised by science. The scientific approach cannot always answer these questions and society as a whole needs to discuss these issues and reach a collective decision.

Often the development and application of science is subject to regulation. You will need to be able to discuss examples of when this happens, for example:

- the role of the regulatory body for UK embryo research
- making decisions about genetic testing on adults and embryos prior to selection for implantation.

Some questions cannot be answered by science, for example, those involving values. You will need to be able to recognise questions that can be answered by using a scientific approach from those that cannot, such as:

- is it right to test embryos for genetic diseases prior to implantation?
- should genetic information about individuals be made available to the police or insurance companies?

Some forms of scientific work have ethical implications that some people will agree with and others will not. When an ethical issue is involved, you need to be able to:

- state clearly what the issue is
- summarise the different views that people might hold.

When discussing ethical issues, common arguments are that:

- the right decision is the one that leads to the best outcome for the majority of the people involved

- certain actions are right or wrong whatever the consequences and wrong actions can never be justified.

You will need to be able to:

- identify examples based on both of these statements
- develop ideas based on both of these statements.

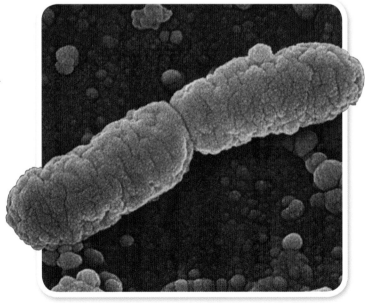

Review Questions

1 Cystic fibrosis is a genetic disorder.

a Choose the **two** words that describe the allele that causes cystic fibrosis:

faulty   **normal**

dominant   **recessive**

b The family tree shows the inheritance of cystic fibrosis.

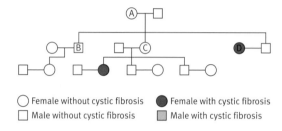

○ Female without cystic fibrosis   ● Female with cystic fibrosis
□ Male without cystic fibrosis   ■ Male with cystic fibrosis

i Which person, **A**, **B**, **C**, or **D**, is a female who has inherited two faulty cystic fibrosis alleles?

ii Which people from **A**, **B**, **C**, and **D**, are carriers?

iii Person **B** has a daughter. We cannot tell from the family tree if the daughter is a carrier. Explain why.

2 Science can show how genetic testing can be carried out. It cannot explain whether it should be carried out.

a Describe some implications of carrying out genetic testing on human beings.

b Explain the ethical issues involved in genetic testing and the different views that might be held.

3 Clones can be produced naturally.

a Which of the examples below are natural clones?

A Two plants made by asexual reproduction from the same parent.

B Two bacteria produced from one bacterium.

C Identical twins.

D Two sperm cells from the same man.

b Clones can look different.

Which factors can cause clones to look different? Choose the correct answer.

genetic factors only

environmental factors only

both genetic and environmental factors

neither genetic nor environmental factors

B2 Keeping healthy

Why study keeping healthy?

Good health is something everyone wants. Stories about keeping healthy are all around you, for example, news reports about what to eat, how much to drink, new viruses, and 'superbugs'. New evidence is reported every day. So the message about how to stay healthy often changes. It's not always easy to know which advice is best.

What you already know

- Microorganisms can enter the body and cause infections.

- White blood cells defend the body against disease.

- Antibiotics can kill some microorganisms but not viruses.

- Immunisation protects against some diseases.

- Some diseases are caused by unhealthy diet and lack of exercise.

- Scientists work together to investigate and reduce the transmission of infectious disease.

Find out about

- how your body fights infections

- arguments people may have about vaccines

- where 'superbugs' come from

- how new vaccines and drugs are developed and tested

- what causes a heart attack

- how scientists can be sure what causes heart disease

- how your body balances water.

The Science

Some diseases are caused by harmful microorganisms. If you are infected your body has amazing ways of fighting back. Vaccines and drugs can help you survive many diseases, and doctors are always trying to develop new ones. But not all diseases are caused by microorganisms. Your lifestyle may also put you at risk of disease. Media reports often warn about the dangers of smoking, eating badly, and not exercising.

Ideas about Science

So, how do you decide which health reports are reliable? Knowing about correlation and cause and peer review will help. There are also ethical questions (arguments about right and wrong) to consider when deciding how we should use vaccines and drugs.

Find out about

- ✔ **how some microorganisms make you ill**
- ✔ **how bacteria reproduce**
- ✔ **infections**

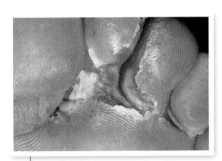

The fungus that causes athlete's foot grows on the skin.

Most days you don't think about your health. It's only when you're ill that you realise how important good health is. Everyone has some health problems during their lives. Usually these are minor – like a cold. But sometimes they can be more serious. Some illnesses may be life-threatening, like heart disease or cancer.

There are lots of reasons for feeling ill. In the doctor's waiting room:

- the man with the painful knee has arthritis
- the young woman feeling sick and tired doesn't know that she's pregnant
- the man having his monthly check-up has had heart disease.

None of these conditions can be passed on to other people. But the other patients all have **infectious** diseases. Infections can be passed from one person to another.

Passing it on

Infections are caused by some **microorganisms** that invade the body. Microorganisms are **viruses**, **bacteria**, and **fungi**.

When disease microorganisms get inside your body, they reproduce very quickly. This causes **symptoms** – the ill feelings you get when you are unwell. Symptoms of infectious diseases can be caused by:

- damage done to your cells when the microorganisms reproduce
- poisons (toxins) made by the microorganisms.

crying, uncomfortable, red gums with white patches on them

cut finger: sore, red cut with pus

swollen glands, runny nose

aching joints, headache, high temperature

monthly check-up

painful, swollen knee joint

sore throat, swollen glands, headache

nausea, tiredness

There are medicines that can cure many diseases caused by bacteria and fungi. But we still don't have many good treatments for diseases caused by viruses. Instead we take medicines that help us feel better until our bodies get rid of the viruses. You will learn more about this later in this chapter.

What are microorganisms like?

Microorganisms are very small. To see bacteria you need a microscope. Viruses are even smaller. They are measured in nanometres, and one nanometre is only one millionth of a millimetre.

Key words

- ✔ **infectious**
- ✔ **microorganisms**
- ✔ **fungi**
- ✔ **viruses**
- ✔ **bacteria**
- ✔ **symptoms**

Microbe attack!

Every breath of air you take has billions of microorganisms in it. And every surface you touch is covered with them. But most of the time you stay fit and healthy. This is because:

- most microorganisms do not cause human diseases
- your body has barriers that keep most microorganisms out.

	Virus	Bacterium	Fungus
Size	20–300 nm	1000–5000 nm	50 000+ nm
Appearance			
Examples of diseases caused	flu, polio, common cold, AIDS, measles	tonsillitis, tuberculosis, plague, cystitis	athlete's foot, thrush, ringworm

Jolene's finger

Jolene cut her finger when she was gardening. She didn't wash it quickly, so bacteria on her skin and in the soil invaded her body. Once inside they started to reproduce. And when bacteria reproduce, they do it in style.

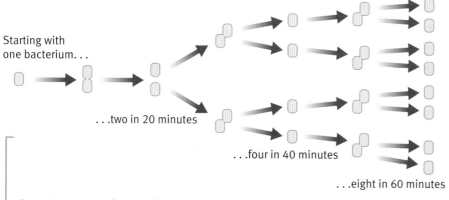

Starting with one bacterium. . .

. . .two in 20 minutes

. . .four in 40 minutes

. . .eight in 60 minutes

Bacteria can reproduce rapidly inside the body.

It was just a small cut, so I ignored it. By the time I went to bed it was a bit sore and red. Now it's all swollen and shiny. It really hurts.

Reproduction in bacteria is simple. Each bacterium splits into two new ones. These grow for a short time before splitting again. If conditions are right – warmth, nutrients, moisture – they can split every 20 minutes.

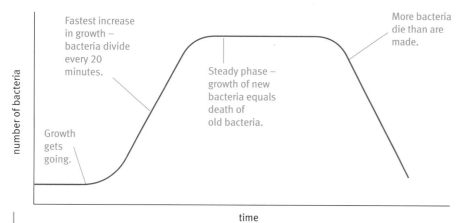

Fastest increase in growth – bacteria divide every 20 minutes.

More bacteria die than are made.

Steady phase – growth of new bacteria equals death of old bacteria.

Growth gets going.

number of bacteria

time

In ideal conditions in a sealed container bacteria can't keep up their fastest growth. Food starts to run out, or waste products kill them off.

Questions

1. Name three types of microorganism that can cause disease.

2. Write down two different diseases caused by each type of microorganism you have named.

3. Explain two ways that microorganisms make you feel ill.

4. What are ideal conditions for bacteria to reproduce?

5. Three harmful bacteria get into a cut. How many might there be after three hours?

Find out about

- ✔ **how white blood cells fight infection**
- ✔ **how you become immune to a disease**

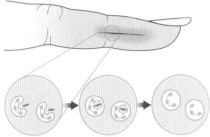

Jolene's body responds by sending more blood to the area.

White blood cells surround the bacteria and digest them.

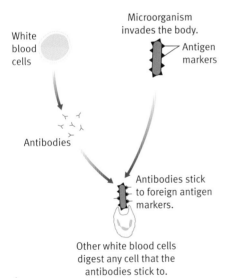

White blood cells

Microorganism invades the body.

Antigen markers

Antibodies

Antibodies stick to foreign antigen markers.

Other white blood cells digest any cell that the antibodies stick to.

One type of white blood cell makes **antibodies** to label microorganisms. A different type digests the microorganisms. All cells have antigen markers on the outside that are unique to that type of cell. The antigen markers on microorganisms are foreign to your body.

The battle for Jolene's finger

Conditions inside Jolene's body are ideal for the bacteria. But they don't have everything their own way.

The redness and swelling in Jolene's finger is called inflammation. Extra blood is being sent to the wounded area, carrying with it the body's main defenders – the **white blood cells**. One type of white blood cell engulfs (surrounds) the bacteria and **digests** them.

The worn-out white blood cells, dead bacteria, and broken cells collect as pus. So redness and pus show that your body is fighting infection. As the bacteria are killed, the inflammation and pus get less until the tissue heals completely.

Your body's army – fighting infection

The parts of your body that fight infections are called your **immune system**. White blood cells are an important part of your immune system.

What's the verdict?

In most cases the body will overcome invading bacteria. Keeping the cut clean and using antiseptic is usually enough treatment. But Jolene's cut is quite deep, so her doctor gives her a course of **antibiotics**. These are antimicrobial chemicals that kill bacteria and fungi but not viruses. Different antibiotics affect different bacteria or fungi.

Everybody needs antibodies – not antibiotics!

A bad cold is something we've all had. And there's not usually much sympathy – 'What's all the fuss about? It's just a cold!'

Natalie has been ill for a few days. Her doctor explains that he won't be giving her any antibiotics. Her cold is caused by a virus, which antibiotics cannot treat. Natalie's own body is fighting the infection by itself.

Fighting the virus

Natalie's neck glands are swollen because millions of new white blood cells are being made there. These white blood cells are fighting the virus in her body.

If antibodies are so good, why do I get ill?

The **antigens** on every microorganism are different. So your body has to make a different antibody for each new kind of microorganism. This takes a few days, so you get ill before your body has destroyed the invaders.

This doesn't really matter for diseases like a cold. But for more serious diseases this is a problem. The disease could kill a person before their body has time to destroy the microorganisms.

Why do you get some diseases just once?

Once your body has made an antibody it can react faster next time. Some of the white blood cells, called **memory cells**, that make the antibody stay in your blood. If the same microorganism invades again, these white blood cells recognise it, reproduce very quickly, and start making the right antibody. This means that the body reacts much faster the second time you meet a particular microorganism. Your body destroys the invaders before they make you feel ill. So you are **immune** to that disease.

Not another one!

Natalie's cold soon got better, but she had only been back at school for about three weeks before she caught another one. If you have an illness like chickenpox, you are very unlikely to catch it again because you are immune. So why do we catch an average of three to five colds every year?

The problem is that there are hundreds of different cold viruses. So every cold you catch is caused by a different virus. To make things worse, the viruses have a very high **mutation** rate. This means that their DNA changes regularly. So do the antigen markers on their surface. The antibody that worked last time will no longer match the marker. Your body needs to make a different antibody to fight the virus. This is why we suffer the symptoms of a cold all over again.

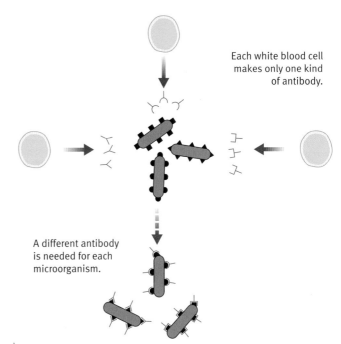

Each white blood cell makes only one kind of antibody.

A different antibody is needed for each microorganism.

Only the correctly shaped antibody can attach to each kind of microrganism.

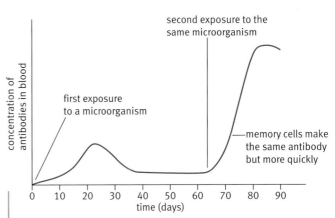

A person is infected twice by a disease microorganism. Their white blood cells make antibodies much faster the second time.

Questions

1 Why are antibiotics not given to patients infected with a virus?

2 Explain two ways that white blood cells protect the body from invading microorganism. You could do this with a diagram.

3 Write down one sentence to describe the job of the immune system.

4 Draw a flowchart to explain how you can become immune to chickenpox.

5 Write a few sentences to explain to Natalie why she will never be immune to catching colds.

Key words
- ✔ **white blood cells**
- ✔ **digests**
- ✔ **immune system**
- ✔ **antibiotics**
- ✔ **antibodies**
- ✔ **antigens**
- ✔ **memory cells**
- ✔ **immune**
- ✔ **mutation**

Find out about

- ✔ **how vaccines work**
- ✔ **deciding if vaccines are safe to use**

In the UK we are lucky to be able to get medicines for many diseases. But it would be even better not to catch a disease in the first place. **Vaccinations** aim to do just that.

Vaccinations make use of the body's own defence system. They kick-start your white blood cells into making antibodies. So you become immune to a disease without having to catch it first.

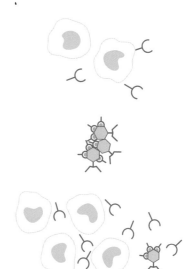

Small amounts of disease microorganisms are put into your body. Dead or inactive forms are used so you don't get the disease itself. Sometimes just parts of the microorganisms are used.

White blood cells recognise the foreign microorganisms. They make the right antibodies to stick to the microorganisms.

The antibodies make the microorganisms clump together. Other white blood cells digest the clump.

Your body stores some of the white blood cells (memory cells). If you meet the real disease microorganism, the antibodies you need are made very quickly.

The microorganisms are destroyed before they can make you ill. (Not to scale)

○ How vaccines work.

Age	Diseases protected against by childhood vaccinations
2, 3 and 4 months	DTB-Hib (diphtheria, tetanus, whooping cough, polio and Hib, a bacterial infection that can cause meningitis and pneumonia) Pneumococcal infection Meningitis C
13 months	MMR (measles, mumps, and rubella)
3–5 years	Diphtheria, tetanus, whooping cough and polio MMR
Girls aged 12–13 years	Cervical cancer caused by human papillomavirus
13–18 years	Diphtheria, tetanus, polio

○ Many childhood diseases are very rare in the UK because of vaccination programmes.

Are vaccines safe?

Any medical treatment you have should do two things:

- improve your health
- be safe to use.

Vaccines can improve your health by protecting you from disease. They are tested to make sure that they are safe to use. But it is important to remember that no action is ever completely safe. People are genetically different, so they react differently to medical treatments, including vaccines.

Doctors decide that a treatment is safe to use when:

- the risk of serious harmful effects is very small
- the benefits outweigh any risk.

You can read more about this in Section E.

Whose choice is it?

To stop a large outbreak of a disease, almost everyone in the population needs to be vaccinated. If they are not, large numbers of the disease-causing microorganisms will be left in infected people. If the vaccination rate drops just a little, lots of people will get ill.

☐ vaccinated   ■ infected   ■ not vaccinated

The vaccination rate is 98%. Unvaccinated people are unlikely to catch the disease.

The vaccination rate has dropped to 90%. Unvaccinated people are much more likely to catch the disease.

Why does the government encourage vaccinations?

Doctors encourage parents to have their children vaccinated at an early age. In the UK there are mass vaccination programmes for some diseases, such as measles. This means that few people suffer from these diseases. Parents have to balance the possible harm from the disease against the risk of possible side-effects from the vaccine.

- Almost nobody who has a vaccine notices any harmful effects.
- Harmful effects from the MMR vaccine can be mild (3 in every 10 000 children), or produce a serious allergic reaction (1 in every million children).
- Some children who catch measles are left severely disabled (1 in every 4000 cases).
- Measles can be fatal (1 in 10 000 cases).

For society as a whole, vaccination is the best choice. But for each parent it is a difficult choice, with their child at the centre of it. People often perceive the risk of vaccination to be greater than the risk of measles. It is important that people have clear and unbiased information to help them make their decision.

Questions

1 What is a vaccine made of?

2 Describe how a vaccine can stop you from catching an infectious disease.

3 Explain why a vaccine can never be 'completely safe'.

In 1998 the MMR vaccine was wrongly linked to autism. This worried many parents. In the 1970s there were similar worries about the whooping cough vaccine.

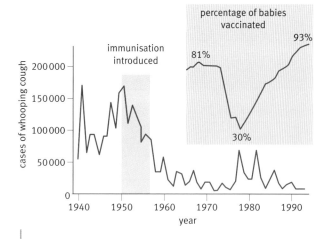

The graph shows the number of whooping cough cases in the UK each year between 1940 and 1992.

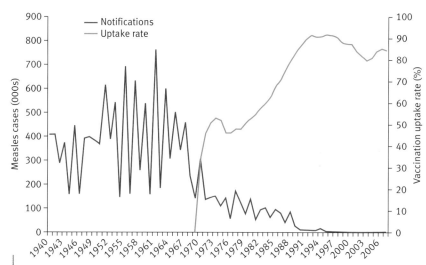

The graph shows how the number of cases of measles changed between 1940 and 2007 and the vaccination uptake since 1970.

Whooping cough:

Disease killed or damaged thousands of children each year.

1950 – vaccine introduced.

1970s – scientific report suggested that there could be a link between the vaccine and serious brain damage in children. Many media stories about the report.

Vaccinations fell from 81% to around 30%.

Over 200 000 extra cases of whooping cough in the 1970s and 1980s, with 100 deaths.

1980s – scientists showed that reports of brain damage from the vaccine had been inaccurate. But it took almost 20 years before vaccinations were back to their original level.

Questions

4 To stop a large outbreak of a disease, almost all of the population must be vaccinated against it. Explain why.

5 a Estimate the number of whooping cough cases one year before vaccination began.
   b Describe what happened to the number of cases between 1950 and 1970.
   c What happened to the percentage of babies vaccinated between 1973 and 1979?
   d Explain why this change happened.

6 Look at the number of whooping cough cases between 1965 and 1990. Is there any link with the percentage of babies vaccinated?

7 The number of cases of measles in England and Wales rose to over 5000 in 2008. Suggest a reason for this.

Smallpox

Smallpox was a devastating disease. In the 1950s there were 50 million cases worldwide. This fell to 10–15 million cases by 1967 because of vaccination by some countries. But 60% of the world's population were still at risk.

In 1967 the World Health Organisation (WHO) began a campaign to wipe out smallpox by vaccinating people across the world. In 1977 the last natural case of smallpox was recorded, in Somalia, Eastern Africa.

Should people be forced to have vaccinations?

Governments and public bodies like the National Health Service make decisions about who should be offered vaccinations based on an asessment of risk and benefit. All children are offered the measles vaccine. There is enough measles vaccine for every child in the UK. If everyone had to be vaccinated by law, there would be a much lower risk of any child catching the disease. However, a few children would still get the disease, because vaccinations don't have a 100% success rate.

So it would be possible for measles vaccination to be compulsory – but it isn't. Society does not think it is right to force anyone to have this particular treatment. There is a difference between what *can* be done with science, and what people think *should* be done.

Different decisions

Where you live may make a difference to your choice about vaccination.

People are more likely to catch a disease due to poor hygiene or overcrowded housing. They will also suffer more if they catch a disease because they may:
- be weaker because of poor diet or other diseases
- have less access to medicines and other healthcare.

So people from poorer communities may make different decisions about vaccinations compared with people in better-off communities.

Smallpox killed every fourth victim. It left most survivors with large scars, and many were also blinded.

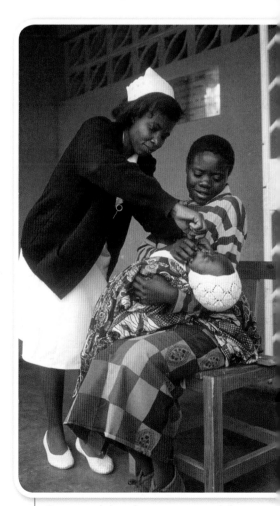

Some people have become concerned about the safety of vaccines for their children. But for many, the decision is easy.

> ## Questions
>
> 8 Explain why measles vaccination is not compulsory in the UK.
>
> 9 Give two reasons why people in different parts of the world may feel differently about having vaccinations.
>
> 10 Scientists cannot make vaccines against every disease. How would you decide which diseases to target?

Find out about

- ✓ where 'superbugs' come from
- ✓ how you can help fight them

Antibiotics are made naturally by bacteria and fungi to destroy other microorganisms. The fungus growing on this bread makes penicillin.

Microorganisms (bacteria, fungi, and viruses) can be killed by antimicrobial chemicals. Some only inhibit their reproduction. The person's immune system destroys those remaining. Antimicrobials used at home, like bleach, will kill, or inhibit, bacteria, fungi, and viruses. Antifungal chemicals kill, or inhibit, fungi. Antibiotics from the doctor only kill bacteria. They are not effective against viruses.

The first antibiotics

The Ancient Egyptians may have been the first people to use antibiotics. They used to put mouldy bread onto infected wounds. Scientists now know that the mould is a fungus that makes penicillin. In the 1940s scientists started to grow the fungus to make larger amounts of penicillin.

The bugs fight back

To begin with, penicillin was called a 'wonder drug'. Before the 1940s, bacterial infections had killed millions of people every year. Now they could be cured by antibiotics. Antibiotics were also used to treat animals. They were even added to animal feed, to stop farm animals from getting infections.

But within ten years, one type of bacteria was no longer killed by penicillin. It had become resistant. New antibiotics were discovered, but each time resistant bacteria soon developed. The 'superbugs' we are dealing with now are resistant to all known antibiotics, except one. How long before bacteria become resistant to it as well, we don't know.

Where have superbugs come from?

- A tiny change in one gene – a mutation – can turn a bacterial cell into a 'superbug'. Just one superbug on its own won't do much

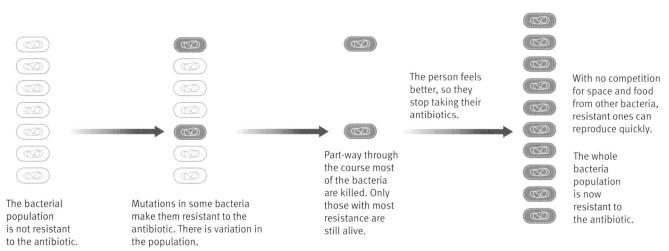

The person feels better, so they stop taking their antibiotics.

With no competition for space and food from other bacteria, resistant ones can reproduce quickly.

The bacterial population is not resistant to the antibiotic.

Mutations in some bacteria make them resistant to the antibiotic. There is variation in the population.

Part-way through the course most of the bacteria are killed. Only those with most resistance are still alive.

The whole bacteria population is now resistant to the antibiotic.

A few mutations can result in antibiotic-resistant bacteria.

damage. But if it reproduces rapidly, it could produce a large population of bacteria, all resistant to an antibiotic. Fungi such as those that cause ringworm and thrush have also become resistant to commonly-used antifungal drugs in exactly the same way.

Why are superbugs developing so quickly?

Two things increase the risk of **antibiotic-resistant** superbugs developing:

- people taking antibiotics they don't really need
- people not finishing their course of antibiotics.

If you are given a course of antibiotics and take them all, it is likely that all the harmful bacteria will be killed. But if you stop taking the antibiotics because you start to feel better, the microorganisms that survive will be those that are most resistant to the antibiotic. They will live to breed another day – and so a population of antibiotic-resistant bacteria soon grows.

How can we stop the superbugs?

Scientists cannot stop antibiotic-resistant bacteria from developing. The mutations that produce these bacteria are part of a natural process. For now, we can only hope that scientists can develop new antibiotics fast enough to keep us one step ahead of the bacteria.

But as well as new drugs, there are other ways of tackling the problem:

- having better hygiene in hospitals to reduce the risk of infection
- only prescribing antibiotics when a person really needs them
- making sure people understand why it is important to finish all their antibiotics (unless side-effects develop).

New drugs in strange places?

Scientists are always looking out for sources of new drugs. For example, crocodile blood might be the source of the next family of antibiotics. A chemical found in crocodile blood is a powerful antibacterial agent. It was discovered by a scientist who wondered why crocodiles didn't die of infections when they bit each other's legs off.

Key word

✓ antibiotic resistant

'SUPERBUGS' MRSA ON THE RAMPAGE

These killer bacteria are resistant to almost all known antibiotics. The bad news is that they have broken out of hospitals. People are dying of MRSA 'superbug' infections picked up at work, out shopping, and even at home. And the cause? The very antibiotics we've been using to kill them!

The bacteria MRSA is resistant to almost all antibiotics.

Crocodile blood could be the source of important new antibacterial drugs.

Questions

1 What are antibiotic-resistant bacteria?

2 Write bullet-point notes to explain how antibiotic-resistant bacteria can develop.

3 Describe two things that you can do to reduce the risk of antibiotic-resistant bacteria developing.

Find out about

- how new drugs are developed
- how they are tested

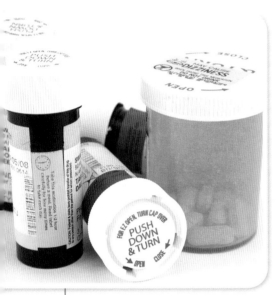

From painkillers to vaccines, antibiotics to antihistamines, medicines are part of everyday life.

Sian is a cancer research scientist.

Most of us take medicines prescribed by our doctor without asking many questions. We assume that they will do us good. But what if we could ask the scientist who developed the medicine some questions?

Is it safe?

How much should I take?

Are there any side-effects?

How did you discover the drug?

Has it been tested properly?

Scientists around the world are trying to develop new drugs: new antibiotics, new treatments for asthma and cancer, and new vaccines for malaria and HIV.

Developing a new drug takes years of research, and lots of money. The rewards for a successful discovery can be huge improvements in human health. For drug companies there may also be large profits.

A scientist explains how a new drug is developed:

First we study the disease to understand how it makes people ill. This helps us work out what we need to treat it – for example, a chemical to kill a microorganism, or a chemical to replace one the body isn't making properly.

We search through many natural sources to find a chemical that may be the correct shape to do this. We look at computer models of the molecules to test our ideas.

When we find a chemical that could work, there are many tests that must be done. It's also important that we could make lots of it without too many problems. Only a very small number of possible drugs get through all these stages.

Stage 1: human cells

Early tests are done on human cells grown in a laboratory. Scientists try out different concentrations of a possible new drug. They test it on different types of body cells with the disease. These tests check how well the chemical works against the disease – how effective it is. They also give the scientists data about how safe the drug is for the cells.

Stage 2: animal tests

If the drug passes tests on human cells, it is tried on animals. Animal trials are carried out to make sure that the drug works as well in whole animals as it does on cells grown in the laboratory.

Stage 3: clinical trials

If the drug passes animal trials then it can be tested on people. These tests are called **human trials** or **clinical trials**. They give scientists more data about the effectiveness and safety of the drug. Scientists carry out human trials on healthy volunteers to test for safety. They then carry out trials on people with the disease to test for effectiveness and safety. Long-term human trials ensure that the drug is safe and works. It is important that the drug is effective and there are no adverse side effects when used for a long time. These studies may continue after a drug is approved for use, providing more data on the safety of the drug.

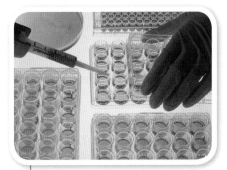

Drugs are tested on cells in the laboratory. These are called in vitro tests.

Trials using animals or human volunteers are called in vivo tests.

Not everybody agrees that it is right to test drugs on animals. The British Medical Association (BMA) believes that animal experimentation is necessary at present to develop a better understanding of diseases and how to treat them, but says that alternative methods should be used whenever possible.

If animal trials go well, we apply for a patent. It costs a lot of money to develop a new drug. If we have a patent, no other company can sell the medicine for 20 years. But because clinical trials and the approval process take many years, we often only have about 10 years when we're the only people making the drug.

Key words
✓ **human trials**
✓ **clinical trials**

Questions

1 Copy and complete the table:

Stage	Testing	To find out
one	Drug is tested on human cells grown in the laboratory.	• how safe the drug is for human cells • how well it works against the disease
two		
three		

2 Developing a new drug is usually very expensive. Suggest why.

Clinical trials – crunch time

Five years ago Anna was diagnosed with breast cancer. Fortunately her treatment worked and she recovered. Now Anna has been asked to take part in the trial of a new drug. Doctors hope it will reduce the risk of the cancer coming back.

What treatment will Anna get?

People who agree to take part in this trial will be put randomly into one of two groups. **Random** assignment into treatment groups is very important in making sure the results of the study are reliable.

One group of people in the trial will be given the new drug, another group will not. This is the **control** group. The results from both groups will be compared.

Anna talks to her doctor:
The problem is I won't know if I'm getting any treatment or not. Could I be risking my health? I know the trial could help people in the future – but what about me? Can you tell me if I will be given the real drug or not?

Before the trial Anna would sign a patient consent form. She signs it to say that all of her questions have been answered. She can also leave the trial at any time. Anyone taking part in a drug trial must give their 'informed consent'.

Anna's doctor wouldn't know if she was getting the new drug or not. Neither would Anna. Someone else would prepare the treatments. This is because Anna would be part of a **double-blind** trial.

If Anna or her doctor knew what treatment she was getting, it could affect the way they report her symptoms. A random double-blind trial is considered the best type of clinical trial.

What treatment will the control group be given?

The drug being tested is a new treatment. In almost all clinical trials the control group are given the treatment that is currently being used. So comparing the results from both groups shows whether the new treatment is an improvement.

Sometimes there isn't any current treatment for an illness. In these cases the control group can be given a **placebo**. This looks exactly like the real treatment but has no drug in it. Using a placebo in a clinical trial is very rare. The control group in Anna's trial will be given a placebo.

Human trials – ethical questions

Taking the placebo would not increase Anna's risk of cancer returning. Taking the new drug may bring other risks. But her doctor will be looking out for any harmful effects. And the new drug may increase her chance of staying well.

It may seem unfair that the control group could miss out on any benefits of the new drug. But remember that not all drugs pass clinical trials. Proper testing is needed to find out if a new drug has real benefits. Tests also give doctors data about the risk of unwanted harmful effects.

- If the trial shows that the risks are too great it will be stopped.
- If the trial shows that the drug has benefits it will immediately be offered to the control group.

Blind trials

In some trials the doctor is told which patients are being given the drug. This may be because they need to look very carefully for certain unwanted harmful effects. The patient still should not know. This method is called a **blind trial**.

Questions

3 Explain why drug trials must be random.

4 Explain the difference between blind and double-blind trials.

5 Describe a situation in which it would be wrong to use placebos in a trial.

6 What do you think Anna should do? Explain why you think this.

double-blind trial

blind trial

open-label trial

Open-label trials

In an **open-label trial** both the patient and the doctor know the treatment. This may be necessary if, for example, a physiotherapy treatment was being compared to a drug treatment, or if a new drug is given to all the patients in a trial. This happens when there is no other treatment and patients are so ill that doctors are sure they will not recover from the illness. The risk of possible harmful effects from the drug is outweighed by the possibility that it could extend their lifespan or be a cure. No one is given a placebo. It would be wrong not to offer the hope of the new drug to all the patients. Penicillin is one example where this happened.

In a drug trial the doctor and/or patient may (✓) or may not (✗) know if the treatment is the new drug.

Key words
- ✓ **random**
- ✓ **control**
- ✓ **blind trial**
- ✓ **placebo**
- ✓ **double-blind trial**
- ✓ **open-label trial**

Find out about

- ✓ **how blood is pumped round your body**
- ✓ **what causes a heart attack**
- ✓ **how to look after your heart**
- ✓ **measuring how hard your heart is working**

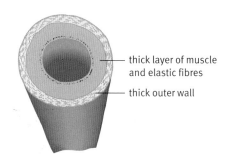

Arteries take blood from the heart to your body. The thick outer walls can withstand the high pressure created by the pumping heart.

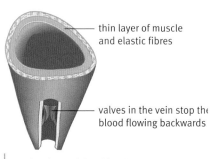

thin layer of muscle and elastic fibres

valves in the vein stop the blood flowing backwards

Veins bring blood back to the heart. The thin layer of muscle and elastic fibres allows the vein to be squashed when you move. This pushes the blood back to the heart.

thin wall (one cell thick) to allow diffusion of oxygen and food to the cells

5–20 μm diameter

Capillaries take blood to and from tissues. The very thin walls (one cell thick) allow oxygen and food to diffuse to cells and waste from cells.

Three weeks ago 45-year-old Oliver suffered a serious heart attack. He was very lucky to survive. Now he wants to try and make sure it doesn't happen again.

Your body's supply route

Your heart is a bag of muscle in your body. When you are sitting down it beats at about 70 beats per minute. It has four chambers. The upper two receive blood and the lower two have thick muscular walls to pump the blood. Your heart is a double pump. Tubes carry the blood around your circulatory system.

How blood circulates

Blood enters the right-hand side of your heart from your body. It flows into the right lower chamber, which pumps it to the lungs to pick up oxygen. Your blood then flows back into the upper chamber on the left-hand side of your heart, then into the left lower chamber. There it is pumped to the rest of your body to deliver oxygen. There are valves between the upper and lower chambers to make sure blood flows in the right direction.

I'll never forget. I went cold and clammy, covered in sweat. And the pain – it wasn't just in my chest. It was down my arm, up my neck and into my jaw. I don't remember much else until I woke up in intensive care. I never want to go through that again.

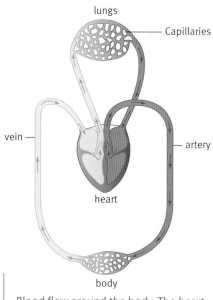

Blood flow around the body. The heart is described as a double pump because the blood passes through the heart twice in each complete circuit.

What is a heart attack?

Blood brings oxygen and food to cells. Cells use these raw materials for a supply of energy. Without energy the heart would stop. So heart muscle cells must have their own blood supply.

Sometimes fat can build up in the coronary arteries. A blood clot can form on the fatty lump. If this blocks an artery, some heart muscle is starved of oxygen. The cells start to die. This is a heart attack.

How serious is the problem of heart disease?

Heart disease is any illness of the heart, for example, a blocked coronary artery and a heart attack.

Oliver survived his heart attack because only a small part of his heart was damaged. He was given treatment to clear the blocked artery. If the blood supply to more of his heart had been blocked, it could have been fatal.

In the UK 230 000 people have a heart attack every year. This is one every two minutes. Coronary heart disease is more common in the UK than in non-industrialised countries. This is because people in the UK do less exercise – most people travel in cars and have machines to do many jobs. And a typical UK diet is high in fat.

What causes heart disease?

Heart attacks are not normally caused by an infection. Your genes, your **lifestyle**, or most likely a mixture of both, all affect whether you suffer a heart attack. There isn't one cause of heart attacks – there are many different **risk factors**. Your own risk of heart disease increases the more of these risk factors you are exposed to.

Is Oliver at risk of another heart attack?

Oliver has a family history of coronary heart disease. He is also overweight, smokes, and often eats high-fat, high-salt food. This diet has given Oliver high blood pressure and high cholesterol levels. All these factors increase his risk of a heart attack. Oliver does like sport – but he'd rather watch it on TV than do exercise himself. Oliver's doctor has given him advice about reducing his risk. This includes advice about drinking alcohol in moderation and reducing his stress levels.

Key words

- ✔ arteries
- ✔ veins
- ✔ capillaries
- ✔ coronary arteries
- ✔ lifestyle
- ✔ risk factors
- ✔ heart disease

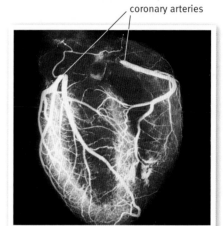

coronary arteries

Coronary arteries carry blood to the heart muscle.

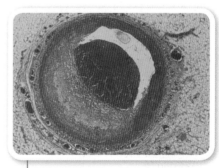

Fat build-up in a coronary artery.

♡ HEALTHY HEART

- Cut down on fatty foods to lower blood cholesterol.
- If you smoke, stop.
- Lose weight to help reduce blood pressure and the strain on your heart.
- Take regular exercise (such as 20 minutes of brisk walking each day) to increase the fitness of the heart.
- Reduce the amount of salt eaten to help lower blood pressure.
- If necessary, take medicines to reduce blood pressure and/or cholesterol level.
- Relax, reduce stress.

Questions

1 Explain why heart cells need a good blood supply.

2 Explain how too much fat in a person's diet can lead to a heart attack.

3 List four lifestyle factors that increase a person's risk of heart disease.

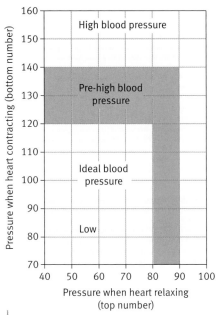

There is a range of 'normal' blood pressure. People differ in height, weight, lifestyle, and sex. All these factors influence what will be normal for that person.

Key words

- ✓ pulse rate
- ✓ blood pressure
- ✓ lifestyle diseases

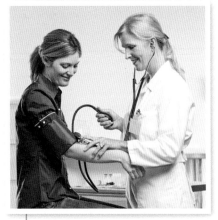

The doctor measures the blood pressure of a patient. High blood pressure can cause heart disease.

Oliver may have been able to prevent his heart attack if he had known his arteries were narrowing. His doctor might have measured his **blood pressure** and **pulse rate**. He could have warned Oliver to live a healthier life.

Monitoring the heart

When arteries become clogged up with fatty deposits, it is more difficult for the blood to be pumped around the body. The heart has to work harder and will have to beat faster.

You can measure how hard your heart is working by measuring your pulse rate. Your pulse is taken on the inside of your wrist or at your neck. You can measure how fast your heart is beating by measuring the beats per minute.

A more accurate measure of how difficult it is for your heart to pump blood around your body is blood pressure. Blood-pressure measurements record the pressure of the blood on the walls of the artery. It is recorded as two numbers, for example, 120/80. The higher number is the pressure when the heart (specifically the left ventrical) has fully contracted and the lower number is when the heart is fully relaxed.

High blood pressure increases the risk of a heart attack. Narrowing of the arteries raises heart rate and blood pressure, and so do drugs such as Ecstasy and cannabis.

Lifestyle diseases

Heart disease and some cancers like lung cancer are **lifestyle diseases**. One hundred years ago infectious diseases killed most people in the UK. Today, better hygiene, vaccinations, and healthcare mean infectious illnesses are more controlled. Lifestyle diseases are much more common than they were. In other parts of the world the lifestyle factors may be different, making other diseases more common.

Questions

4   Heart disease is more common in the UK than in non-industrialised countries. Suggest why.

5   Your neighbour wants to do more exercise, but she gets bored easily and doesn't want to spend money going to the gym. Suggest some ways she could get exercise into her daily life.

Causes of disease – how do we know?

It's usually easy for doctors to find the cause of infectious diseases. The microorganism is always in the patient's body. It is harder to find the causes of lifestyle diseases, like heart disease or cancer.

Health warning in 1971. Health warning in 2003.

Smoking and lung cancer

Government health warnings have been printed on cigarette packets since 1971. There was evidence showing a link – a **correlation** – between smoking and lung cancer. But in 2003 the message was made much stronger. How did doctors prove that smoking *caused* lung cancer?

An early clue

In 1948 a medical student in the USA, Ernst Wynder, observed the autopsy of a man who had died of lung cancer. He noticed that the man's lungs were blackened. There was no evidence that the man had been exposed to air pollution from his work. But his wife told Wynder that he had smoked 40 cigarettes a day for 30 years. Wynder knew that one case is not enough to show a link between any two things.

In 1950, two British scientists, Richard Doll and Austin Bradford Hill, started a series of scientific studies. First, they compared people admitted to hospital with lung cancer to another group of people in hospital for other reasons. Smoking was very common at the time, so there were lots of smokers in both groups. But the percentage of smokers in the lung cancer group was much greater.

This data showed a link – a correlation – between smoking and lung cancer. Doll and Hill suggested smoking caused lung cancer. But, a correlation doesn't always mean that one thing causes another.

Find out about

✔ how scientists identify risk factors for lifestyle diseases
✔ the evidence needed to prove a causal link

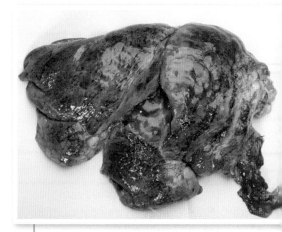

Lung tissue blackened by tar from cigarette smoke.

Cigarettes smoked per day	Number of cases of cancer per 100 000 men
0 – 5	15
6 – 10	40
11 – 15	65
16 – 20	145
21 – 25	160
26 – 30	300
31 – 35	360
36 – 40	415

The data shows how the number of cases of lung cancer in men is affected by the number of cigarettes smoked.

How reliable was the claim?

Doll and Hill published their results in a medical journal so that other scientists could look at them. This is called 'peer review'. Other scientists look at the data and how it was gathered. They look for faults. If they can't find them, then the claim is more reliable.

The claim is also more reliable if other scientists can produce data that suggests the same conclusions.

A major study

In 1951 Doll and Hill started a much larger study. They followed the health of more than 40 000 British doctors for over 50 years. The results were published in 2004 by Doll and another scientist, Richard Peto. They showed that:
* smokers die on average 10 years younger than non-smokers
* stopping smoking at any age reduces this risk

The last piece of the puzzle – an explanation

Lung cancer rates in the USA rose sharply after 1920. The same pattern was seen in the UK.

Many doctors were now convinced that smoking caused lung cancer. But cigarette companies did not agree. They said other factors could have caused the increase in lung cancer, for example, more air pollution from motor vehicles.

The missing piece of the puzzle was an explanation of *how* smoking caused cancer. In 1998 scientists discovered just this. They were able to explain *how* chemicals in cigarette smoke damage cells in the lung, causing cancer. This confirmed that smoking *causes* cancer.

Questions

1 Write down one example of an everyday correlation.

2 Draw a graph to show how the number of cases of lung cancer in men is affected by the number of cigarettes smoked.

3 Explain briefly what happens during 'peer review'.

4 Explain why scientists think it is important that a scientific claim can be repeated by other scientists.

5 It's unlikely that many people would have agreed with Wynder if he'd reported the case he saw in 1948. Suggest two reasons why.

6 If a man smokes 20 cigarettes a day from age 16 to 60, will he definitely develop lung cancer? Explain your answer.

Before 1920 lung cancer was very rare. As smoking became more popular with men, the numbers of lung cancer cases rose. This happened later for women, because very few women smoked until after World War II.

What makes a good study?

There are many reports in the media about studies of health risks. These **epidemiological studies** look for diseases caused by different risk factors. For example, what are the key risk factors for heart disease?

You may want to use this information to make a decision about your own health. So it's important to know if the study has been done well. There are several things you can look for.

How many people were involved in the study?

A good study usually looks at a large sample of people. This means that the results are less likely to be affected by chance.

In the USA a long-term study looked at over 13 000 people across three generations. This study has been hugely important for heart disease research. It has led to the identification of all known major risk factors for heart disease.

How well matched are the people in the study?

Health studies sometimes compare two groups of people. One group has the risk factor, the other doesn't, for example, a study that compares people who exercise with people who do not. In these studies it is important to **match** the people in the two groups as closely as possible.

Genetic studies of heart disease

A large-scale **genetic study** by the Wellcome Trust Case Control Consortium was published in 2007. It showed that heart disease is due to genes and lifestyle.

The research team studied the genomes (types of genes a person has) of 2000 people with coronary heart disease and 3000 healthy controls. They indentified six common alleles or gene variants that are associated with heart disease.

Understanding the genetics that lead to heart disease will help to tell us how much risk a person faces. A person who carries one or more of the 'risk' alleles can still reduce their risk by adopting a healthy lifestyle, monitoring their blood pressure and cholesterol levels, and taking medication.

How big is the risk?

There's one other thing to look for when using data from health studies to make decisions. Imagine a headline like 'Risk of disease is doubled'. It's important to check how big the original risk is. For example, what if the risk of an outcome is that it will happen to one person in a million? An increase of two times is still only two in one million – or one in every 500 000 people. This is still a very small risk.

Looking at the health of lots of people can show scientists the risk factors for different diseases.

Key words

- ✓ **epidemiological studies**
- ✓ **genetic studies**
- ✓ **match**

Questions

7 Name one factor that increases a person's risk of heart disease.

8 Suggest two things you should look for when deciding whether a study was well planned.

9 Your teenage daughter has started smoking. She says 'I don't believe smoking causes heart disease or lung cancer. Grandad has smoked all his life, and he's fine.' How would you explain to her that she may not be so lucky?

Find out about

- ✔ homeostasis
- ✔ why it is important
- ✔ negative feedback

Inside your cells thousands of chemical reactions are happening every second. These reactions are keeping you alive. But for your cells to work properly they need certain conditions. Keeping conditions inside your body the same is called **homeostasis**.

Homeostasis is not easy – lots of things have to happen for your body to 'stay the same'. Look at just a few of the changes happening every second.

Your body works hard to:

- keep the correct levels of water and salt
- control the amounts of nutrients
- get rid of toxic waste products, for example, carbon dioxide and urea.

Control systems

The control systems keeping a steady state in your body work in a similar way to artificial control systems.

All control systems have:

- a **receptor**, which detects the stimuli (the change)
- a **processing centre**, which receives the information and coordinates a response
- an **effector**, which produces an automatic response.

Running has made this athlete hot. His body is sweating more to cool back down. This is an example of homeostasis.

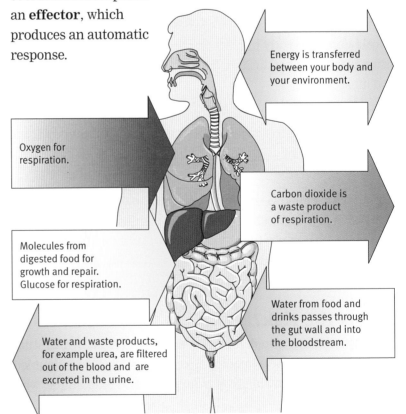

Energy is transferred between your body and your environment.

Oxygen for respiration.

Carbon dioxide is a waste product of respiration.

Molecules from digested food for growth and repair. Glucose for respiration.

Water from food and drinks passes through the gut wall and into the bloodstream.

Water and waste products, for example urea, are filtered out of the blood and are excreted in the urine.

Some of the inputs and outputs that are going on all the time in your body.

How does an incubator work?

Premature babies cannot control their temperature, so they are put in incubators. The incubator is an artificial control system.

An incubator has a temperature sensor, a thermostat with a switch, and a heater. If the temperature in an incubator falls too low, the heater is switched on. The temperature goes up. When the temperature is high enough, the heater is switched off. This type of control is called **negative feedback**:

• any change in the system results in an action that reverses the change.

The diagram below explains how this works.

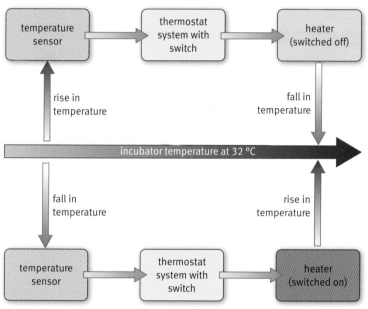

Negative feedback in an incubator to control temperature

An artificial control system is keeping this baby's temperature steady.

Negative feedback systems are all around you. For example, if the temperature inside your fridge goes up, the motor switches on to cool it down. When it is cool enough, the motor switches off.

What about your body?

Some of the temperature control in your body is automatic too. For example, you do not consciously decide to sweat when you are hot. In the same way, water control is automatic. You do not decide to feel thirsty or to make less urine. These changes are coordinated by both nervous and hormonal communication systems.

Your body also uses negative feedback systems, although your body's systems are more complicated than in the incubator. You have effectors to cool you down and other effectors to warm you up. Negative feedback systems reverse any change to the system's steady state.

Questions

1 Write down a definition for homeostasis.

2 In an incubator, name:
   a a receptor
   b a processing centre
   c an effector.

3 Write down a definition for negative feedback.

Water homeostasis

Find out about

- ✓ **how your body gains and loses water**
- ✓ **how your kidneys get rid of waste**
- ✓ **how kidneys balance your water level**

Experiments carried out with students showed that being able to drink water in the classroom
- increased their concentration time
- improved test results.

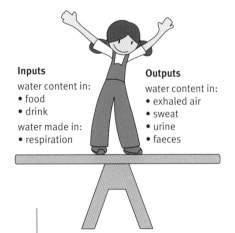

Inputs
water content in:
- food
- drink

water made in:
- respiration

Outputs
water content in:
- exhaled air
- sweat
- urine
- faeces

Water intake and water loss must balance for your body to work well.

Water homeostasis, keeping a steady water level, is done by balancing your body's water inputs and water outputs. The diagram on the opposite page shows how you gain and lose water.

What do your kidneys do?

Your kidneys have two jobs: water homeostasis and **excretion**. Excretion is getting rid of toxic waste products from chemical reactions in your cells. These two jobs are linked because you use water to flush out waste products.

Your **kidneys** control the water balance in your body. They do this by changing the amount of urine that you make. On a hot day, or when you have been running, you lose a lot of water in sweat. So your kidneys make a smaller volume of urine but with the same amount of waste. Your urine will be more concentrated that day.

Getting the water level in cells right is important to maintain the correct chemical concentration levels for cell activity.

More about water balance

Remember that the concentration and volume of your urine varies. On cold days you probably make lots of pale-coloured urine. On hot days you make a smaller volume of darker, more concentrated urine.

The concentration of your blood plasma determines how much water your kidneys reabsorb, and how much you excrete in urine. The concentration of your blood can become higher than normal because of:
- excess sweating because of increased exercise levels
- not drinking enough water
- eating salty food.

In these cases your kidneys will reabsorb more water, making less urine.

Drugs and urine

Some drugs affect the amount of urine a person makes. **Alcohol** causes a greater volume of dilute urine to be produced and can make people very dehydrated. Dehydration can cause dizziness, headaches, and tiredness. Ongoing dehydration can have adverse effects on health including problems with your kidneys, liver, joints, and muscles. Severe dehydration can cause low blood pressure, seizures, increased heart rate, and loss of conciousness.

The drug **Ecstasy** has the opposite effect. It reduces the volume of urine a person makes. It also affects the body's temperature control. Overheating may lead to the person drinking too much water. The amount of water in the body can become dangerously high, causing seizures that can be fatal. Ecstasy also, increases blood pressure and heart rate, increasing the risk of a heart attack.

Controlling water balance

The control system for water balance is a negative feedback system.

- Receptors in the brain detect any changes in concentration in the blood plasma.
- When the concentration is too high, it triggers the release of a hormone called **ADH** from the **pituitary gland** in the brain. When the concentration is low, no ADH is released.
- The ADH travels in the blood to the kidneys. These are the effectors. ADH affects the amount of water that can be reabsorbed back into the blood. The more ADH, the more water is reabsorbed.

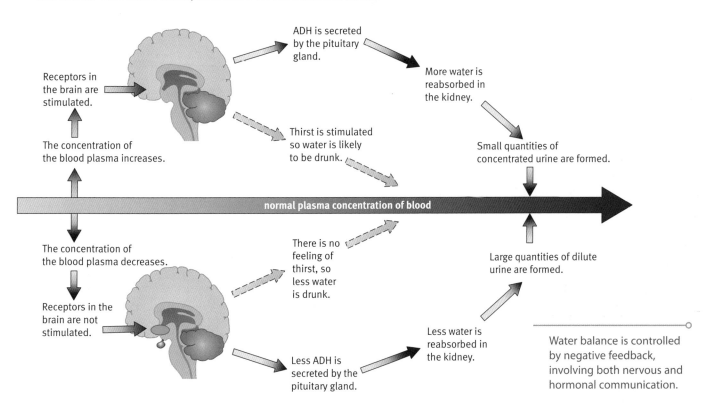

Water balance is controlled by negative feedback, involving both nervous and hormonal communication.

Drugs affect ADH control

Alcohol and Ecstasy change the volume of urine a person makes because they affect ADH production. Alcohol suppresses ADH production. Less water is reabsorbed in the kidneys, so a larger volume of urine is made. Ecstasy increases ADH production, resulting in a smaller volume of urine as more water is reabsorbed in the kidneys. Body-fluid build-up can lead to brain damage and death.

Questions

1 Copy and complete the table below to show what happens when the concentration of the blood plasma changes.

	Concentration of blood plasma falls	Concentration of blood plasma rises
Pituitary gland secretes	less ADH	
Kidney reabsorbs		
Urine volume		
Urine concentration		increases

2 Ecstasy triggers release of ADH. Explain the effect this will have on water balance.

Science Explanations

Keeping healthy involves maintaining a healthy lifestyle, avoiding infection, using medication when necessary, and our bodies maintaining a constant internal environment.

You should know:

- about microorganisms multiplying rapidly in the human body, damaging cells and releasing toxins to produce disease symptoms
- how white blood cells engulf and digest microorganisms or produce antibodies to destroy them
- how specific antibodies recognise different microorganisms
- how memory cells provide the body with immunity by making specific antibodies rapidly if the body is re-infected, destroying the microorganisms
- how a safe form of a microorganism, called a vaccine, causes the body to produce antibodies
- that drugs and vaccines can never be completely safe because people are different genetically and react differently
- that antimicrobials are used to kill bacteria, fungi, and viruses
- how mutations in microorganisms can make them resistant to antimicrobials
- how new drugs are tested on animals, human cells, and healthy volunteers, then ill people for safety and effectiveness
- about long-term drug trials, including open-label, blind, and double-blind trials
- why using placebos in human trials raises ethical issues
- that the circulatory system consists of the heart, which is a double pump with its own blood supply, arteries, veins, and capillaries
- how fatty deposits in blood vessels can trigger heart disease
- how genetic and lifestyle factors, such as diet, exercise, stress, smoking, and misuse of drugs, can cause heart disease
- how to measure a person's pulse rate
- that high blood pressure may indicate heart disease
- how nervous and hormonal systems help maintain homeostasis
- how body systems detect stimuli with receptors, coordinate responses with a processing centre, and produce a response with effectors
- how negative feedback between receptors and effectors helps to maintain homeostasis
- how kidneys regulate water balance by producing dilute or concentrated urine
- how the hormone ADH controls urine concentration through negative feedback and how drugs affect ADH.

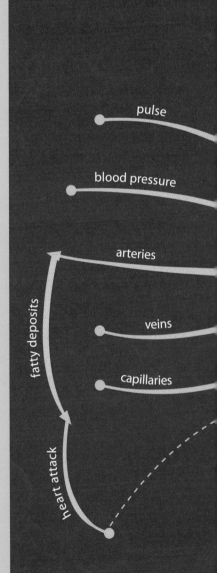

pulse

blood pressure

arteries

veins

capillaries

fatty deposits

heart attack

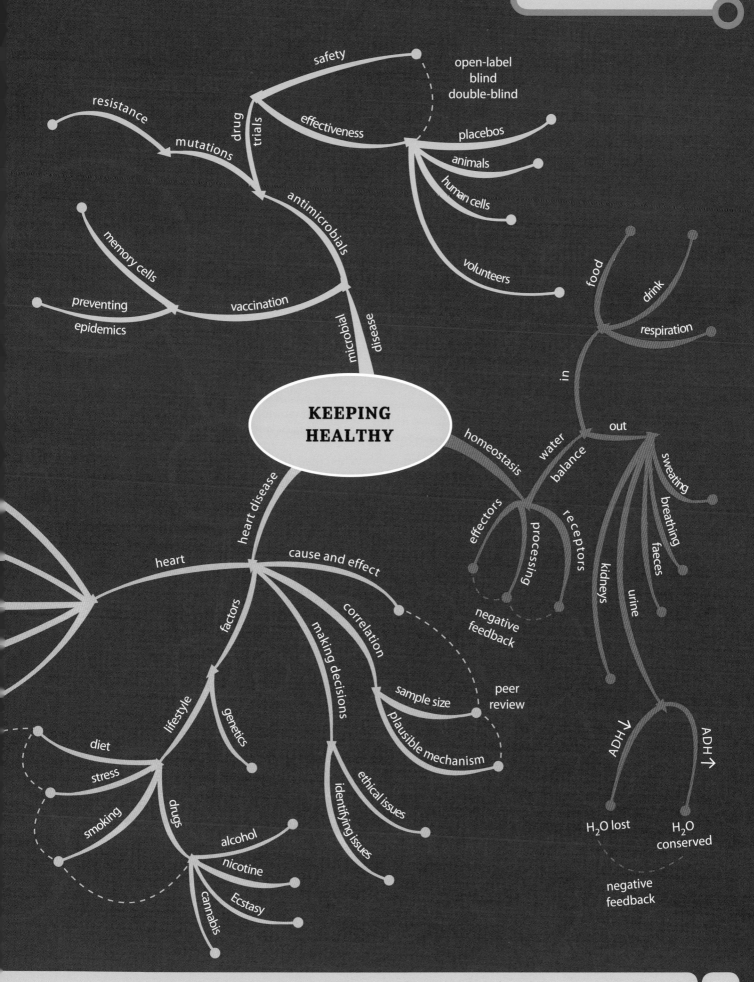

safety

open-label
blind
double-blind

resistance

drug

mutations

trials

effectiveness

placebos

animals

human cells

antimicrobials

volunteers

memory cells

preventing

epidemics

vaccination

microbial

disease

food

drink

respiration

in

**KEEPING
HEALTHY**

homeostasis

water

out

balance

sweating

effectors

processing

receptors

breathing

heart disease

faeces

negative
feedback

kidneys

urine

heart

cause and effect

factors

correlation

sample size

peer
review

making decisions

plausible mechanism

lifestyle

genetics

ADH

ADH

diet

stress

ethical issues

smoking

drugs

identifying issues

alcohol

nicotine

H_2O lost

H_2O
conserved

cannabis

Ecstasy

negative
feedback

Ideas about Science

This module provides opportunities to develop your understanding of cause–effect explanations, how scientists share their ideas, and how decisions about scientific issues are made, including ideas about risk.

If an outcome increases or decreases as an input variable increases there is a correlation between the two.

You should be able to:
- suggest and explain an example of a correlation from everyday life, such as an increase in the number of cigarettes smoked increases the risk of developing heart disease
- identify a correlation when given data such as text, a graph, or a table
- understand that a correlation does not prove a cause and that the outcome might be caused by some other factor, for example, icecream sales increase as hayfever increases, but icecream does not cause hayfever.

Scientists investigate claims that a factor increases the probability of an outcome, such as the link between cigarette smoking and heart disease, by closely matching different groups of the population or choosing them randomly. You should be able to critically evaluate such studies by commenting on sample size and how well the samples are selected or matched.

Even when evidence exists that a factor is correlated to an outcome, scientists look for a causal mechanism. For example, smoking increases the effect of heart disease because of the effects of nicotine on the body. Nicotine is the mechanism.

Scientists report all their claims to scientific conferences or scientific journals. This is so other scientists can peer review the evidence and claims.

This gives the claim credibility, especially when the findings have been replicated by another scientist.

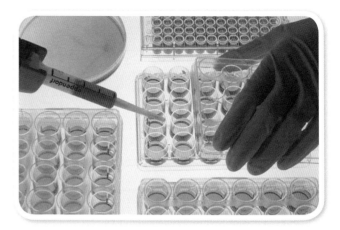

Some questions cannot be answered by science, for example, those involving values. You will need to be able to distinguish questions that can be answered by using a scientific approach from those that cannot, such as should vaccinations be compulsory?

When discussing these questions, the benefits and the size of the perceived and measured risk must be considered.

Some forms of scientific work have ethical implications that some people will agree with and others will not. When an ethical issue is involved, you need to be able to:
- state clearly what the issue is
- summarise the different views that people might hold.

When discussing ethical issues, common arguments are that:
- the right decision is the one that leads to the best outcome for the majority of the people involved
- certain actions are right or wrong whatever the consequences, and wrong actions can never be justified.

You will need to be able to:
- identify examples based on both of these statements.

Review Questions

1 Our bodies are sometimes invaded by microorganisms. We can protect ourselves by having a vaccination containing dead microorganisms. Some of the statements below describe how vaccination helps to protect us from disease-causing microorganisms.

Use only the correct statements and place them in the correct order.

a The disease-causing microorganisms are destroyed.

b White blood cells produce lots of antigens.

c Memory cells rapidly make antibodies for the disease.

d The body slowly makes antibodies to the disease.

e The disease-causing microorganisms enter the body.

f The disease releases antibodies into the blood.

g We receive a vaccination against the disease.

2 New drugs are tested for effectiveness and safety. Explain what and who the drugs are tested on and the role played by open-label, blind, and double-blind trials.

3 Doctors have to make decisions about when to use placebos and who they should be used on. Explain what placebos are, why they are used, and when they should not be used.

4 Kidneys regulate the amount of water present in the human body.

a Explain the role played by ADH in the functioning of the kidneys and how ADH secretion is controlled by negative feedback.

b Explain the effects of:

i alcohol on the functioning of the kidney

ii Ecstasy on the functioning of the kidney.

5 Eating a diet containing a lot of fatty food can increase the risk of heart disease.

Different people have different views about this.

To answer these questions, you may use each person once, more than once, or not at all.

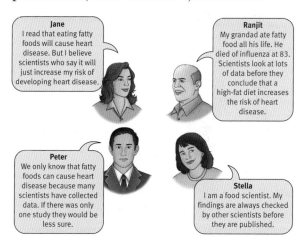

Jane
I read that eating fatty foods will cause heart disease. But I believe scientists who say it will just increase my risk of developing heart disease.

Ranjit
My grandad ate fatty food all his life. He died of influenza at 83. Scientists look at lots of data before they conclude that a high-fat diet increases the risk of heart disease.

Peter
We only know that fatty foods can cause heart disease because many scientists have collected data. If there was only one study they would be less sure.

Stella
I am a food scientist. My findings are always checked by other scientists before they are published.

a Which person says that the absence of replication is a reason for questioning a scientific claim?

b Which person is suggesting that individual cases do not provide convincing evidence for or against a correlation?

c Which person is describing the process of peer review?

d Which **two** people are suggesting that factors might increase the chance of an outcome but not always lead to it?

B3 Life on Earth

Why study life on Earth?

There are over 30 million species of living things on Earth today. Where do they all come from? Why is there so much variety? Is that variety important? Can we learn to look after life on Earth better for future generations? These are the big questions we ask science to answer. Scientists think life began on Earth 3500 million years ago. The first simple organisms have developed and changed, and many species have become extinct.

What you already know

- Characteristics are determined by genetic information passed on from both parents.

- Individuals with the same genetic information may vary.

- Some characteristics are influenced by environmental factors.

- Photosynthesis is the source of biomass in plants.

- Plants need mineral salts for growth.

- The distribution of organisms in an environment is affected by environmental factors.

- Organisms only survive in a habitat where they have all the essentials for life and reproduction.

- Organisms show adaptations to environmental conditions.

- Food webs show feeding relationships.

Find out about

- how different species depend on each other

- how life on Earth is evolving

- how scientists developed an explanation for evolution

- why some species become extinct, and whether it matters.

The Science

Fossils and DNA provide evidence for how life on Earth has evolved. Simple organisms gradually change to form new species.

All life forms depend on their environment and on other species for their survival. Ultimately all life depends on the Sun's energy, and on nutrients, which are recycled through the environment.

Ideas about Science

Today most scientists agree that evolution happens. But nobody thought this 200 years ago. Developing new explanations takes evidence and imagination. Even then, the explanation will change as we collect new evidence and test the ideas. The variety of species on Earth is a valuable resource, which humans depend on. Scientists can help us devise ways to use natural resources in a more sustainable way.

Find out about

- ✓ why living things are all different
- ✓ what a species is
- ✓ adaptations of organisms

Human skin cells and cells in these butterfly wings use the same chemical reaction to make pigment.

You can usually see the differences between different kinds of living things on Earth. But there are also a lot of similarities, even between living things that don't look the same. For example, almost all living things use DNA to pass on information from one generation to the next.

Classification – working out where we belong

Scientists use the similarities and differences between living things to put them into groups. You've probably come across this idea before. It's called **classification**. The biggest group that humans belong to is the kingdom *Animalia* (the Animal Kingdom). The smallest is *Homo sapiens*, or human beings. *Homo sapiens* is our **species** name.

Classification names are in Latin, so that everyone can use the same name for something. It doesn't matter what languages two people speak, they can always use the same Latin name.

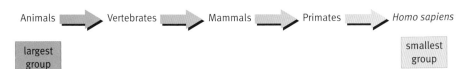

Animals ➡ Vertebrates ➡ Mammals ➡ Primates ➡ *Homo sapiens*

largest group | smallest group

You are most closely related to other members of *Homo sapiens*. But you belong to these other groups as well. Vertebrates are all the animals that have skeletons with backbones. All mammals have five-digit limbs. Primates have five-fingered hands and feet, shoulder joints that can move in all directions, and forward-facing eyes in bone sockets. Down the classification, the groups contain fewer organisms that have more and more characteristics in common.

What makes a species?

Scientists define a species as a group of organisms so similar that:

- they can breed together
- their offspring can also breed (they are **fertile**).

horse

donkey

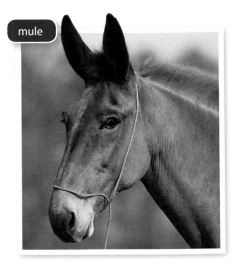

mule

Horses and donkeys do look very similar. But their offspring are infertile. So horses and donkeys are different species.

Horses and donkeys are good examples to explain species. They can breed together and produce offspring called mules. But mules are **infertile**. Horses and donkeys look pretty similar, but they are different species.

All members of a species are not the same. You only need to look around your classroom to see that this is true. This variation is caused by a mixture of genes and environment. It is very important in evolution. You'll find out more about this later.

The art of survival

All of the 30 million or so species presently alive on Earth are successful survivors. They have features that help them survive in their environments. These features are called **adaptations**.

Cactus plants live in hot, dry, desert conditions. They survive because they are adapted to this environment. They store water inside their stems and have large root systems to get water from deep in the soil. Many cactus plants have hard spines instead of leaves.

Fish have special adaptations that enable them to live in water. A fish absorbs oxygen dissolved in water. The oxygen diffuses from the water into the fish's blood across the large surface area of the gills. A streamlined body and a smooth surface helps the fish move through the water with little resistance. Sets of fins keep the fish balanced in the water. A swim bladder gives the fish buoyancy in the water.

Cacti and fish are both adapted to the habitat where they live.

Questions

1 What species do you belong to?

2 Explain why horses and donkeys are different species.

3 Give a reason for each of the cactus's adaptations.

Key words

- ✔ species
- ✔ fertile
- ✔ infertile
- ✔ adaptations

Find out about

- ✓ why some species become extinct
- ✓ how organisms are interdependent

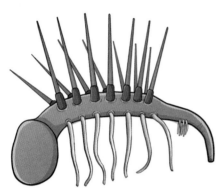

Fossil from burgess shale. This amazing-looking species lived 505 million years ago. It is extinct but scientists think it may be the ancestor of crabs and centipedes.

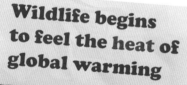

Wildlife begins to feel the heat of global warming

Six regions were studied, representing 20% of the Earth's land area.

A large international study says that up to a quarter of the species on Earth face extinction from global warming.

Over the past few million years, many species of plants and animals have lived on Earth. Most of these species have died out. When all the members of a species die out it is **extinct**.

There is fossil evidence of at least five mass extinctions on Earth. Now we seem to be at the beginning of another.

Around the world over 12 000 species of plants and animals are at risk of extinction. They are endangered.

Where an animal or plant lives is called its **habitat**. Any quick changes in their habitat can put them at risk of extinction.

Changes in the environment

All living things need factors like water and the right temperature to survive. Rising temperatures are changing many habitats. This global warming is putting many species at risk.

New species

New species moving into the habitat can put another species at risk of extinction.

- Animals and plants compete with each other for the things they need. Two different species that need exactly the same things cannot live together. One wins the **competition** for resources like food and shelter.
- The new species could be a **predator** of the species already living there.
- If the new species causes **disease**, it could wipe out the native population.

Red squirrels used to live all over the UK. Now the larger American grey squirrels have taken over most of their habitats.

In the 1960s, the virus that causes Dutch Elm disease came to the UK. It destroyed most of the UK elm population.

Going hungry

Plants and animals need other species in their habitat. For example, in this food chain spiders eat caterpillars.

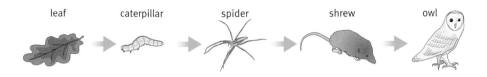

So if the caterpillars all died, the spiders could be at risk. That could also endanger the shrew and the owl.

The food web

Most animals eat more than one thing. Many different food chains contain the same animals. They can be joined together into a **food web**. This shows the **interdependence** of living organisms.

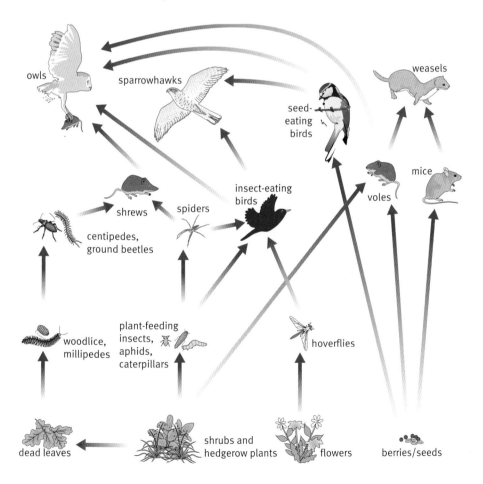

A new animal coming into a food web can affect plants, animals, and microorganisms already living there. The loss of one organism will also affect other organisms in the food web.

Questions

1 Look at the food web on this page.
   a Name two different animals competing for the same food source.
   b A disease kills all the flowering plants. Explain what happens to the number of hoverflies.
   c Mink move into the habitat. They eat voles.
      i The number of mice decreases. Explain why.
      ii Explain what would happen to the number of caterpillars.

2 Explain what is meant by extinct.

3 Name two things that:
   a plant species may compete for
   b animal species may compete for.

Find out about

- ✔ **how organisms depend on the Sun's energy**
- ✔ **how energy and nutrients pass through food webs**

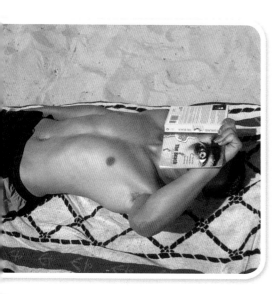

The sunbather enjoys the warmth of the Sun, but does not rely directly on the Sun's energy for his life processes. The plant is harnessing light energy to drive food production.

Nearly all organisms are ultimately dependent on energy from the Sun. The Sun provides energy to keep the Earth's atmosphere warm and drive the production of food chemicals.

Food chains like the one on page 79 all follow the same pattern. They start with plants, the **producers**.

Plants capture energy from sunlight. They use the energy to build organic compounds such as glucose from carbon dioxide and water. This is the process of **photosynthesis**.

Energy from sunlight is stored in these new compounds, which make up the plants' cells. The compounds can be broken down in **respiration** to release energy. The energy stored is passed to other organisms as the plants are eaten or decompose.

Plants trap only about 1–3% of the light energy that reaches their leaves in new plant material. This might sound small, but remember that the Sun's energy output is enormous. So this 1–3% is still enough energy to power life on Earth.

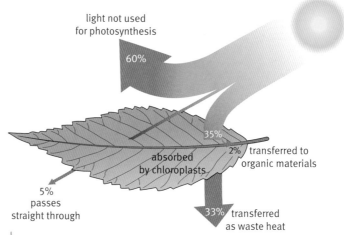

light not used for photosynthesis

60%

35%

absorbed by chloroplasts

2% transferred to organic materials

5% passes straight through

33% transferred as waste heat

Most of the light energy reaching a leaf is reflected from the surface, is transferred as waste heat, or passes straight through the leaf. Only a small percentage is absorbed by the chloroplasts where photosynthesis takes place.

Energy transfer

Animals have to eat. They cannot make their own food so they need to take in organic molecules. They are **consumers**, and they break down food molecules in respiration.

Some of the energy released by respiration is used for growth, where the food molecules become part of the structure of new cells. Animals also use energy released by respiration for other life processes, for example, keeping warm.

On average only about 10% of the energy at each stage of a food chain gets passed on to the next level. The rest:

- is used for life processes in the organism, such as movement
- escapes into the environment as heat energy
- is excreted as waste and passes to decomposers
- cannot be eaten and passes to decomposers

This means that the number of organisms usually gets smaller at each level of an ecosystem. The food chain is limited to only a few levels. Usually there are only the producers, primary consumers, secondary consumers, and tertiary consumers.

The diagram opposite shows the efficiency of energy flow through an ecosystem.

Energy also flows into **decomposers** and **detritivores** as they feed on dead organisms and waste material in the ecosystem. Bacteria and fungi are decomposers. Detritivores are small animals like woodlice.

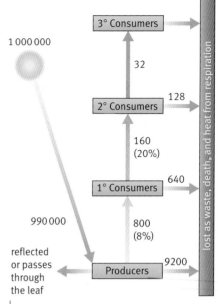

Energy flow through an ecosystem.

Woodlice, earthworms, millipedes, and insect larvae (maggots) are examples of detritivores.

Bacteria and fungi are decomposers. This photograph shows fungus growing on a dead weevil.

Key words

- ✓ **producers**
- ✓ **photosynthesis**
- ✓ **consumers**
- ✓ **respiration**
- ✓ **decomposers**
- ✓ **detritivores**

Questions

1 Look at the food web on page 79. Identify two:
   a producers
   b primary consumers
   c secondary consumers
   d tertiary consumers.

2 Explain why the energy in a producer will not all be transferred to the primary consumer in the same food chain.

3 Calculate the percentage efficiency of energy transfer between the secondary and tertiary consumers in the ecosystem above (shown in the diagram at the top of the page).

Find out about

- ✔ how carbon and nitrogen are recycled through the environment
- ✔ how environmental change can be measured using living and non-living indicators

Key words

- ✔ **combustion**
- ✔ **carbon cycle**
- ✔ **phytoplankton**
- ✔ **photosynthesis**
- ✔ **respiration**
- ✔ **decomposition**

Questions

1 List ways in which carbon dioxide is:
   a added to the atmosphere
   b taken out of the atmosphere.

2 How is human activity causing the amount of carbon dioxide in the atmosphere to rise?

3 Before 1970 the species of phytoplankton called *Ceratium trichoceros* was only found off the south coast of England. It is now found in the waters off Scotland. Suggest what environmental change is likely to have occurred to affect where this phytoplankton lives.

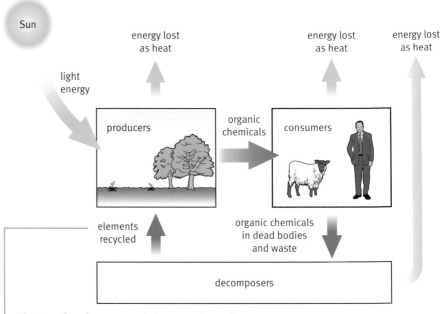

The transfer of energy and elements through an ecosystem.

Energy stored in the chemicals making up cells is passed on to other organisms along food chains. In this way energy moves through the ecosystem. This is similar to how elements like carbon and nitrogen move through the ecosystem. But there is a big difference.

Carbon and nitrogen are always being recycled in an ecosystem.

Recycling carbon

There is only a certain amount of carbon on Earth. Much of the carbon is in molecules that make up the bodies of living things. A lot is also in the atmosphere and oceans as carbon dioxide, and in molecules of fossil fuels.

Carbon dioxide is taken out of the atmosphere by **photosynthesis**. The carbon is used to produce glucose molecules. The glucose is broken down during **respiration**. This releases carbon dioxide back into the atmosphere.

When an animal or plant dies its organic compounds are broken down by microorganisms; this is called **decomposition**. The carbon and nitrogen atoms become part of a new organism in the system.

Carbon dioxide is also added to the atmosphere by the **combustion** (burning) of wood and fossil fuels, such as oil, gas, coal, and petrol. The stages in recycling carbon on Earth are shown in the diagram of the **carbon cycle** on the opposite page.

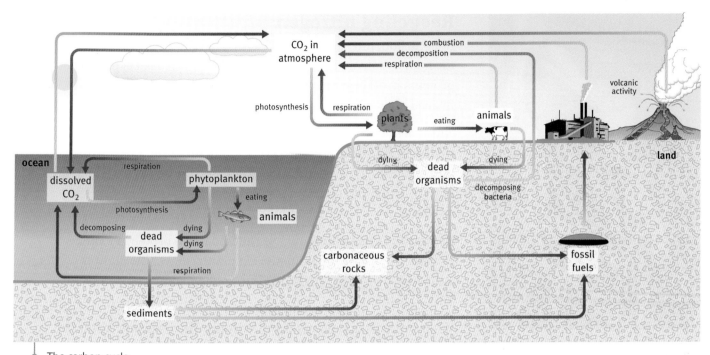

The carbon cycle.

Measuring environmental change

If the amount of carbon dioxide released into the air does not balance the amount taken up by photosynthesis, atmospheric carbon dioxide levels change.

Most scientists agree that the average level of carbon dioxide in the atmosphere is rising. The official figure in 2010 was 0.04% carbon dioxide. It is expected to be 0.05% by the end of the century. The current value is about 40% higher than any value measured over the past 800 000 years.

It is thought that the rise in carbon dioxide levels is linked to rises in the global temperature of the Earth. This is called global warming.

This climate change can be measured by looking at the impact on living organisms. **Phytoplankton** are tiny floating plants found in seawater. Their numbers and patterns of distribution are affected by the impact of climate change on water temperature, ocean mixing, and nutrients levels in the water.

Many species are threatened with extinction if conditions change beyond their ability to adapt.

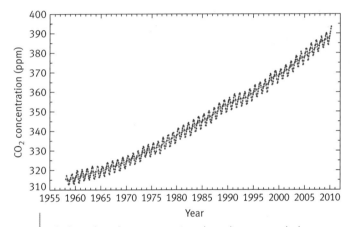

Carbon dioxide concentrations have been recorded at Mauna Loa in Hawaii since 1958. They rise and fall each year, but the overall trend has been an increase of about 1.5 ppm per year since 1980.

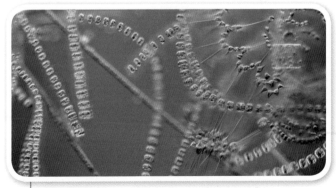

The distribution and abundance of species of phytoplankton (small drifting aquatic plants) change as the temperature of surface water rises.

Recycling nitrogen

Like carbon, nitrogen is recycled through the environment, passing through the air, soil, and living organisms. Nitrogen is part of the proteins in all organisms. Plants cannot take up nitrogen from the air. Instead, they take up nitrogen compounds, such as nitrates, from the soil.

Nitrates in the soil are made using nitrogen from the air. This is called **nitrogen fixation**. **Nitrogen-fixing bacteria** can turn nitrogen gas into nitrates. These bacteria live in the soil and inside swellings on the roots of leguminous plants, such as peas, beans, and clover. The swellings are called root nodules.

Plants take up nitrates and use them to make proteins. Primary consumers (herbivore animals) eat plants, digest the plant proteins, and use the products (called amino acids) to make their own animal proteins.

When animals and plants die they decay. **Decomposer bacteria** break down the proteins in dead organisms. This releases nitrates back into the soil so they are available for plants to absorb once again. Nitrogen in waste from excretion (urine and faeces) is also recycled in this way.

Nitrogen-fixing bacteria live in root nodules on these bean roots.

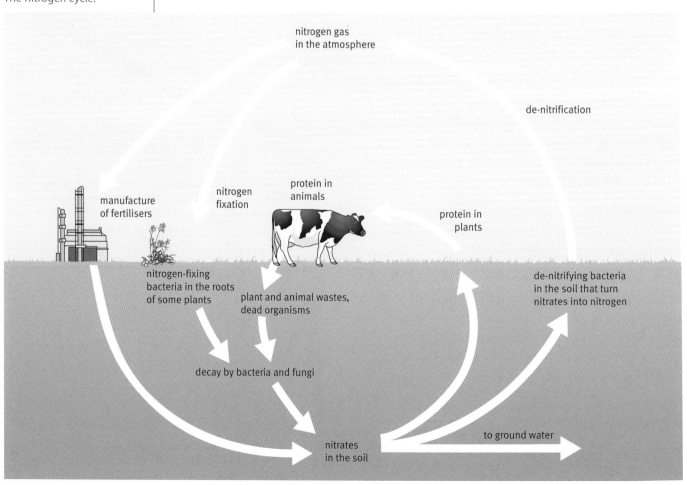

The nitrogen cycle.

nitrogen gas in the atmosphere

de-nitrification

manufacture of fertilisers

nitrogen fixation

protein in animals

protein in plants

nitrogen-fixing bacteria in the roots of some plants

plant and animal wastes, dead organisms

de-nitrifying bacteria in the soil that turn nitrates into nitrogen

decay by bacteria and fungi

to ground water

nitrates in the soil

Denitrifying bacteria break down nitrates in the soil and release nitrogen gas back into the air. This process is known as **denitrification**. All the steps that make up the **nitrogen cycle** are shown in the diagram opposite.

Measuring environmental change

Farmers often add nitrogen to their fields in the form of chemical fertiliser (often ammonium nitrate). This helps make the soil more fertile and increases plant growth. But this can also have a negative impact, causing environmental change.

Chemical fertiliser is very soluble in water. Water rich in nitrate can drain from fields into streams, rivers, and lakes, raising nitrate levels in the water. High levels can cause the rapid growth of microscopic organisms (called plankton), form in massive 'algal blooms'.

Sometimes these blooms can cause sickness to people or animals drinking the water. Decomposer bacteria break down dead plankton. The bacteria respire using up oxygen dissolved in the water. Animals that need high levels of oxygen, such as **mayfly larvae**, cannot survive in water that contains high levels of nitrates.

Measuring nitrate levels and monitoring invertebrates allows changes in water quality to be measured.

Monitoring air quality

Nitrogen compounds can also pollute the air. Air quality can be monitored by studying the types of lichens surviving in a particular area.

Lichens are unusual organisms made up of a fungus growing with a green alga. The alga provides nutrients for the fungus through photosynthesis, while the fungus provides protection for the alga from drought and UV light from the Sun.

Some species of lichen can grow in areas with high levels of nitrogen compounds in the air, while others are very sensitive to nitrogen compounds and will only grow in places with no air pollution. Sensitive species are usually more feathery types. They become locally extinct when air pollution rises.

Key words

- ✓ nitrogen fixation
- ✓ nitrogen-fixing bacteria
- ✓ decomposer bacteria
- ✓ denitrifying bacteria
- ✓ denitrification
- ✓ nitrogen cycle
- ✓ mayfly larvae
- ✓ lichens

Monitoring invertebrates like these mayfly larvae, also known as mayfly nymphs, allow changes in water quality to be measured.

The golden shield lichen can live in areas with high levels of nitrogen, especially ammonia.

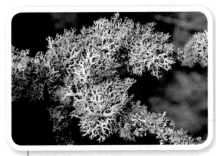

The heather rags lichen is very sensitive to nitrogen pollution in the air, so is rarely found close to roads in big cities.

Questions

4 Explain how nitrogen gets from the air into plants.

5 A gardener notices that the feathery-type lichens have disappeared from her garden. What change in the environment may have occurred?

6 Make a table summarising the different types of bacteria involved in the nitrogen cycle and stating the role of each.

Explaining similarities – the evidence for evolution

Scientists agree that life on Earth began about 3500 million years ago. Life started from a few simple living things. This explains why living things have so many similarities.

These simple living things changed over time to produce the enormous variety of different species of living things on Earth today. The changes also produced many species that are now extinct. This process of change is called **evolution**, and it is still happening today.

What evidence is there for evolution?

Fossils are the remains of dead bodies of living things preserved in rocks. They are very important as evidence for evolution. Almost all fossils found are of extinct species. This is more than 99% of all species that have ever lived on Earth.

How reliable is fossil evidence?

Conditions have to be just right for fossils to develop. Only a very few living things end up as fossils. So there are gaps in the fossil record.

Scientists have collected millions of fossils. This huge amount of evidence has helped to build up a picture of evolution. For example, scientists examined fossils of a dinosaur called *Sinosauropteryx* and discovered orange and white rings down its tail, made of primitive feathers. This provided evidence that the first birds developed from small meat-eating dinosaurs.

Find out about

- ✓ how fossils and DNA provide evidence for evolution

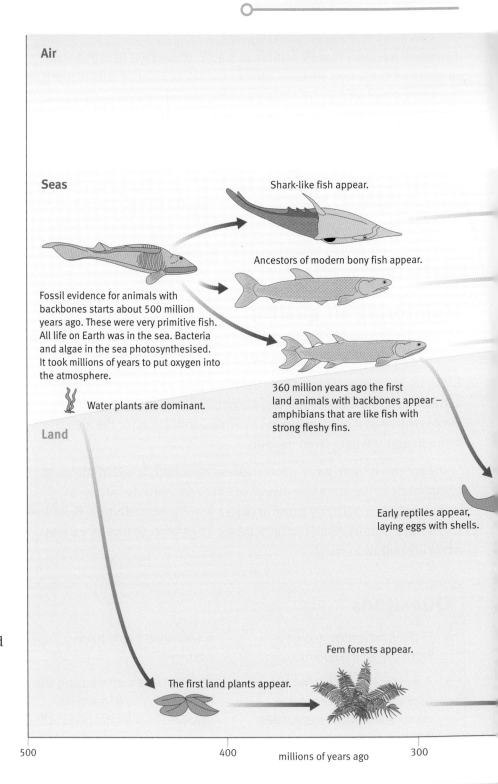

Air

Seas

Shark-like fish appear.

Ancestors of modern bony fish appear.

Fossil evidence for animals with backbones starts about 500 million years ago. These were very primitive fish. All life on Earth was in the sea. Bacteria and algae in the sea photosynthesised. It took millions of years to put oxygen into the atmosphere.

Water plants are dominant.

360 million years ago the first land animals with backbones appear – amphibians that are like fish with strong fleshy fins.

Land

Early reptiles appear, laying eggs with shells.

Fern forests appear.

The first land plants appear.

500   400   300

millions of years ago

What other evidence do we have for evolution?

Scientists also compare the DNA from different living things. The more similar the DNA of two living things, the more closely related they are. This helps scientists to classify them, work out where different species fit on the evolutionary tree, and make sense of the variety of species.

Key words
- ✓ evolution
- ✓ fossils

Humans and chimpanzees have over 98% of their DNA the same. They were thought to share a common ancestor about 6 million years ago. They are both classified as Primates.

flying reptiles

EXTINCTION

modern birds

modern sharks

small meat-eating dinosaur

modern bony fish

most to EXTINCTION except for modern birds

modern amphibians

dinosaurs

modern reptiles

modern horse

early horses

mammals

early humans

Modern people are dominant.

Conifers and ferns are dominant.

Flowering plants appear.

Flowering plants are dominant.

Humans and mice are both mammals. About 85% of their DNA is the same. They shared a common ancestor about 75 million years ago.

200   100   today

millions of years ago

Questions

1 What percentage of all life on Earth is alive now?

2 Name two types of evidence that scientists use as evidence for evolution.

3 Describe the changes that have occurred in the evolution of modern birds.

Evolution did not just happen in the past. Scientists can measure changes in species that are happening now. They expect evolution will continue in the future. Humans can alter the evolution of some species.

Selective breeding

Early farmers noticed that there were differences between individuals of the same species. They chose the crop plants or animals that had the features they wanted, for example, the biggest yield or the most resistance to diseases. These were the ones they used for breeding. This way of causing change in a species is called **selective breeding**. It has been used for breeding wheat, sheep, dogs, roses, and many other species.

Natural selection

People have been using poisons to kill head lice for many years. In the 1980s, doctors were sure that **populations** of head lice in the UK would soon be wiped out.

But a few headlice survived the poisons. These lice bred and now parts of the country are fighting populations of 'superlice' that are resistant to the poisons.

So headlice are another example of change. But this wasn't selective breeding – no one *wanted* to cause superlice. Something in the environment caused the change. This is called **natural selection**. Natural selection is how evolution happens.

Head lice are changing because of human beings. But humans haven't been around on Earth for very long. Most changes to species happened before human beings arrived. Something else in the environment caused the change.

Find out about

- ✓ **how evolution happens – natural selection**
- ✓ **how humans have changed some species**

Selective breeding has produced tulips with different coloured flowers.

Head lice are quite common. They feed on blood.

Key words

- ✓ **selective breeding**
- ✓ **populations**
- ✓ **natural selection**

For many years people used the same shampoo to kill head lice.

A few head lice in the population were able to survive. Their cells were probably able to break down the poison.

'Superlouse' was more likely to breed than the head lice killed by the poison.

Eggs laid by 'Superlouse' hatched into lice that also survived the poison.

These lice spread to other people and bred.

The number of resistant lice in the population increased. People couldn't get rid of their head lice.

Scientists developed a new poison to kill the head lice.

The cycle began again – and the species changed a little more.

Steps in natural selection

① *Living things in a species are not identical. They have variation.*

Ancestors of modern giraffes had variation in the length of their necks.

② *They compete for things like food, shelter, and a mate. But what if something in the environment changes?*

Food supply became scarce. The giraffes competed for food.

③ *Some will have features that help them to survive. They are more likely to breed. They pass their genes on to their offspring.*

Taller giraffes were able to eat more food, so were more likely to survive and breed. They passed on their features to the next generation.

④ *More of the next generation have the useful feature. If the environment stays the same, even more of the following generation will have the useful feature.*

Over many generations, more giraffes with longer necks were born. The number of taller giraffes in populations increase.

Treating HEAD LICE

Your Local Health Authority issues a directive, known as a rotational policy, every two to three years to inform people about which type of insecticide is currently recommended for use in your area.

The rotational policy is intended to prevent head lice becoming resistant to treatment – in other words, to help ensure that the treatments available continue to be effective in killing lice.

Questions

1   How does evolution happen?

2   Copy and complete the table below to compare selective breeding and natural selection.

Steps in selective breeding	Steps in natural selection
Living things in a species are not all the same.	Living things in a species are not all the same.
Humans choose the individuals with the feature that they want.	
These are the plants or animals that are allowed to breed.	
They pass their genes on to their offspring.	
More of the next generation will have the chosen feature.	
If people keep choosing the same feature, even more of the following generation will have it.	

3   Explain what is meant by a population.

4   Read the extract from a leaflet about head lice. Explain how this rotational policy stops the evolution of resistant populations of head lice.

5   Natural selection is sometimes described as 'survival of the fittest'. How good a description of natural selection do you think this is?

Find out about

- ✓ **how Darwin explained evolution**
- ✓ **how explanations get accepted**

Today most scientists agree that evolution happens. But evolution wasn't always as well accepted. A very important person in the story of evolution was Charles Darwin. His ideas were a breakthrough in persuading people that evolution happens.

Darwin's big idea

Darwin worked out how evolution could happen. He explained how natural selection could produce evolution. But he didn't come up with this idea overnight. It took many years.

Charles Darwin was born in 1809. He was interested in plants and animals from a young age. When he was 22, Darwin was given the chance to sail on HMS *Beagle*. The ship was on a five-year, round-the-world trip to make maps.

Journey of the *Beagle*

The *Beagle* stopped at lots of places along the way. At each stop Darwin looked at different types of animals and plants. He collected many specimens and made lots of observations about what he saw. He recorded this data in notes and pictures.

One place the *Beagle* stopped at was the Galápagos Islands, near South America. As he travelled between the different islands, Darwin noticed variation in the wildlife. One thing Darwin wrote about after his trip was the different species of finches living on the Galápagos Islands.

Darwin on HMS *Beagle*.

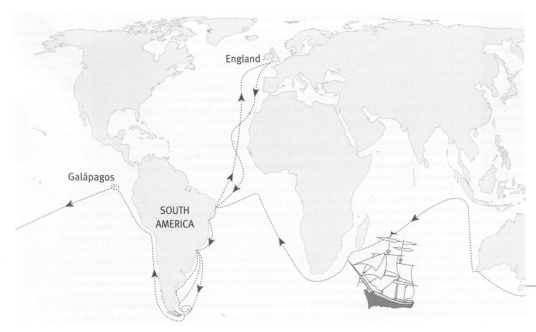

The *Beagle* stopped at different places around the world.

The famous Galápagos finches

Each species of finch seemed to have a beak ideally suited for eating particular things. For example, one had a beak like a parrot for cracking nuts. Another had a very tiny beak for eating seeds. It was as if the beaks were adapted to eating the food on each different island.

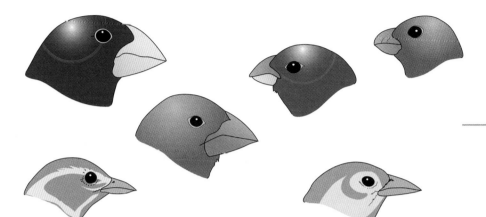

Different species of finch.

In his notes, Darwin started to ask himself a question. He wondered if all the different finches could have evolved from just one species.

What was special about Darwin?

Darwin wasn't the first scientist to think that evolution happens. His own grandfather was one of several people who had written about it earlier. But most people at the time didn't agree with evolution. Darwin was the first person to make a strong enough argument to change their minds.

He started by looking at lots of living things. He made many observations that he would use as evidence for his argument.

- He thought about the evidence in a way that no-one had done before. He was more creative and imaginative.
- He came up with an idea to explain *how* evolution could happen – natural selection.

Darwin showed his notes to a friend, Thomas Huxley. Huxley was also a scientist. When he read them, Huxley said: "How stupid of me not to have thought of this first!"

> One might really fancy that from an original paucity of birds in this archipelago one species had been taken and modified for different ends.

Charles Darwin, *The Voyage of the Beagle*, 1839.

> I look to the future to young and rising naturalists who will be able to view both sides of the question with impartiality.

Charles Darwin, *On the Origin of Species*, 1859.

Questions

1 Darwin made many observations about different species. How did he record his data?

2 What personal qualities did Darwin show that helped him develop his explanation of natural selection?

More evidence back home

Back in England, Darwin moved to a new home in Kent. For 20 years he worked on his idea of natural selection. He exchanged letters with other scientists in different parts of the world. All the time, Darwin was looking for more evidence to support his ideas.

His new home, Down House, had some pet pigeons. They had many different shapes and colours. But Darwin knew they all belonged to the same species. So he realised that:

- animals or plants from the same species are all different – there is **variation.**

Darwin found more evidence for natural selection at home.

Too many to survive

Next, Darwin realised that:

- there are always too many of any species to survive.

He came to this conclusion after reading the work of a famous economist, Thomas Malthus. At the end of the 18th century, Britain's population was growing very fast. Malthus pointed out that the numbers of any species had the potential to increase faster than any increase in their food supply. He predicted that the human population would grow too large for its food supply, and that poverty, starvation, and war would follow.

All the plants or animals of one species are in **competition** for food and space. A lot of them don't survive.

Darwin put these ideas together. He saw that some animals in a population were better suited to survive than others. They would breed and pass on their features to the next generation. This natural selection could make a species change over time. Darwin had explained how evolution could happen.

Elephants usually reproduce from age 30–90. Darwin worked out that after 750 years, if they all survived, there would be nearly 19 million elephants from just one pair!

Owing to this struggle for life, any variation, however slight, if it be in any way profitable to an individual of any species, will tend to the preservation of that individual, and will generally be inherited by its offspring. I have called this principle, by which each variation, if useful, is preserved, by the term of Natural Selection.

Charles Darwin, *On the Origin of Species*, 1859.

Key words
- ✔ variation
- ✔ competition

Same data, different explanations

Other scientists also saw that living things were different. They also saw fossils that showed changes in species. Fifty years before Darwin published his ideas, a French scientist called Lamarck had written a different explanation to Darwin's. He said that animals changed during their lifetime. Then they passed these changes on to their young. He used the example of a giraffe, as shown opposite.

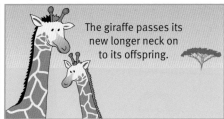

Giraffe evolution explained by Lamarck.

Why was Darwin's explanation better?

A good explanation does two things:
* It accounts for all the observations.
* It explains a link between things that people hadn't thought of before.

Lamarck's explanation said that 'nature' had started with simple living things. At each generation, these got more complicated. If this kept happening, simple living things, like single-celled animals, should disappear. So his idea didn't account for some observations, for example, why simple living things still existed on Earth.

Darwin's idea could better account for these observations. It also linked together variation and competition, which hadn't been done before.

Lamarck's ideas may sound a bit daft now, but he was a good scientist trying to explain changes in species.

Why was Darwin worried about his explanation?

Darwin was worried about how people would react. He wrote his idea of natural selection into a book. Then he wrapped the manuscript in brown paper and stuffed it in a cupboard under the stairs. He wrote a note for his wife explaining how to publish the manuscript when he died. It stayed in the cupboard for almost 15 years.

Questions

3 How did Darwin try to get more evidence to support his ideas?

4 What two things make a good explanation?

5 What two things did Darwin link together to work out his explanation of natural selection?

I never saw a more striking coincidence. If Wallace had my manuscript sketch written out he could not have made a better abstract!

Charles Darwin, in a letter to the geologist Charles Lyell.

It's a disgrace – the thought of us being related to apes!

God made every animal and plant unique. He put fossils on Earth to show us his many designs.

People agreed with Darwin's observations. But they didn't agree with his explanation.

At the 1860 British Association for the Advancement of Science (BA) meeting, Huxley and Hooker argued in favour of Darwin's theory.

On the Origin of Species

Then, in 1856, Darwin received a letter from another scientist, Alfred Russell Wallace. In it Wallace wrote about the idea of natural selection. Darwin was stunned. He gave Wallace credit for what he had done, and the two of them published a short report of some of their ideas. But now Darwin wanted to publish his full book before Wallace, or anyone else, beat him to it.

The now famous *On the Origin of Species* was published in November 1859. This book caused one of the biggest arguments in the history of science.

Many people in Victorian England disagreed with the idea of evolution of life by natural selection.

They were unhappy about the idea that humans were related to apes.

Why did people start to believe in evolution?

The British Association for the Advancement of Science meets every year. Scientists meet to share their ideas. In 1860, many scientists argued against Darwin's idea.

But two of his friends, Thomas Huxley and Joseph Hooker, defended it. They were very good scientists. They were also very good at speaking in public. So they helped to change many people's minds about natural selection.

The end of the story?

Natural selection was a good explanation. However, there were three big problems with it. But it wasn't Darwin's opponents who spotted these. It was Darwin himself.

Firstly, he knew that the record of fossils in the rocks was incomplete. At that time it was even more difficult to trace changes from one species to another than it is today. New fossil evidence has been found since then to support the idea of natural selection.

Secondly, the age of the Earth had not been worked out accurately enough. In Darwin's time it was thought to be about 6000 years old. So there didn't seem to have been enough time for evolution of complex organisms to have taken place. Scientists now have evidence from studying decay of radioactive atoms in rocks to show that the Earth is about 4.5 billion years old.

The last problem was in two parts:

- Darwin could not explain why all the living things in one species were not all the same. Where did variation come from?

- Also, he could not explain how living things passed features on from one generation to the next.

Both of these puzzles would have been easier for Darwin to answer if he had known about genes. Scientific discoveries since his time have allowed other scientists to do this.

At the same time that Darwin was writing *On the Origin of Species*, an Austrian monk called Gregor Mendel (1822–84), was breeding pea plants. From his experiments he discovered dominant and recessive alleles – different versions of the same genes. Mendel's work explained how features were passed on. He sent a copy of his work to Darwin, but Darwin didn't realise how important it was. Mendel's work was largely ignored until 16 years after his own death.

The debate goes on

In 1996, the late Pope John Paul II, head of the Roman Catholic Church, acknowledged Darwin's ideas with the words: "… new scientific knowledge leads us to recognise more in the theory of evolution than hypothesis."

People continue to debate evolution. Because many of them have strong personal beliefs that are affected by this idea, the debate is unlikely to stop anytime soon.

Questions

6 Most people in the 1800s disagreed with natural selection. What evidence did they have against this explanation?

7 Do you agree that evolution happens? Explain why you think this.

8 Suggest why scientists are sometimes reluctant to give up an accepted explanation, even when new data seems to show it is wrong.

Find out about

✓ **how new species are formed**

A mutation in a gene controlling fur colour produced tigers with white fur.

What we need here is a bit of variation!

Questions

1 Explain what a mutation is, and how it can happen.

2 What four processes combine to produce a new species?

Key words
✓ **mutation**
✓ **reproductive isolation**

Charles Darwin's theory of evolution by natural selection predicts that new species will be formed from existing species and that other species will become extinct. These events usually happen slowly over many generations, which is why Darwin was not able to observe them happening. Since Darwin's time scientists have learnt a lot about DNA, and this has helped them to understand how new species form.

Species show variation

We saw earlier that a species is a group of organisms that can breed together to produce fertile offspring. They cannot reproduce successfully with members of different species. All the members of a species are not identical – there is variation.

Mutations cause variation

Suppose that, when DNA is being copied, a mistake is made. This **mutation** could result in a different coloured flower, or spots on an animal's fur. Mutations happen naturally, and they are also caused by some chemicals or ionising radiation.

Mutations produce differences in a species. They are a cause of variation. This is very important for natural selection. Without variation, natural selection could not take place.

Most mutations have no effect on the plant or animal. Mutations that do have an effect are usually harmful. Very, very rarely a mutation causes a change that makes an organism better at surviving. If the mutation is in the organism's sex cells, it can be passed on to its offspring.

Living in an uncertain world

If the environment changes then only some of the population will survive. By natural selection, only individuals with features that make them adapted to the new environment will survive.

Living in splendid isolation

Populations that are isolated from each other have no contact with their neighbours. Organisms will be able to reproduce with other members of their own population, but will never meet organisms from other populations.

Sometimes variation might arise in one population that will prevent the organisms reproducing successfully with those from neighbouring populations, even if they were able to meet. This is called **reproductive isolation**. The isolated population has become a new species.

Variety of life

It has been estimated that there could be over 30 million species on Earth. This huge variety of different animals, plants, fungi, algae, and microorganism species on Earth and the genetic variation within them is called **biodiversity**.

If new species are being formed, does it matter that some become extinct? Isn't extinction just part of life? Twenty-First Century Science put this question to Georgina Mace of the UK Zoological Society.

"It is true that species have always gone extinct. This is a natural process. But the pattern of extinction today is different from what has been recorded in the past.
- The rate of species extinction today is thousands of times higher than in the past.
- Current extinctions are almost all due to humans."

Georgina Mace

Find out about

- ✔ why it matters if species become extinct
- ✔ biodiversity and sustainability

Are humans to blame for some extinctions?

In 1598, Dutch sailors arrived on the island of Mauritius in the Indian Ocean. In the wooded areas along the coast they found fat, flightless birds that they called dodos. By 1700, all the dodos were dead. The species had become extinct. The popular belief is that sailors ate them all. But this explanation appears too simple. Written reports from the time suggest that they were not very nice to eat.

What killed the dodos?

Humans may not have eaten dodos. But did they cause their extinction without meaning to? When the sailors arrived, they brought with them rats, cats, and dogs. These may have attacked the dodos' chicks or eaten their eggs. The sailors also cut down trees to make space for their houses. Maybe this took away the dodos' habitat.

So human beings can cause other species to become extinct:
- directly, for example by hunting
- indirectly, for example by taking away their habitat or bringing other species into the habitat.

Dodos were not able to survive the changes in their environment. This is a disaster for any species.

Foxgloves are very poisonous. But they have given us a powerful medicine to treat heart disease.

Monoculture crop production helps to maximise yields and profits but reduces biodiversity.

The number of supermarket plastic bags used is falling, but the total packaging of food and other goods is rising. Large-scale packaging use is not sustainable.

Does extinction matter?

If many species become extinct, there will be less variety on Earth. This variety is very important. For example:

- People depend on other species for many things. Food, fuel, and natural fibres (such as cotton and wool) all come from other species. Some of today's food crops have been developed from wild plants using selective breeding. The wild relatives can still be found, although they are very different to the domesticated varieties. For example, wild potatoes are poisonous and wild sugar cane produces very little sugar. Wild plants will continue to be used for selective breeding to develop new crops.
- Many medicines have come from wild plants and animals. There are probably many other medicines in plants that haven't been found yet.
- Ecosystems that have high biodiversity tend to cope more easily with natural disasters. In a drought, a species with lots of genetic variation is more likely to survive because some individuals will be better adapted to the drier conditions. Some species will not survive at all, becoming locally extinct. But if there are lots of different species in the ecosystem, then some should survive.

Biodiversity and sustainability

Keeping biodiversity is part of using Earth in a sustainable way. **Sustainability** means meeting the needs of people today without damaging Earth for people of the future. The conservation of different species is important if the Earth is going to be a good home for future generations.

Farmers have to grow the food we need to eat but in a sustainable way that will not damage the environment. Farmers prefer to grow crops in large fields, which can be harvested by machines. A single crop of genetically uniform plants will grow at the same rate and be ready to harvest at the same time. This is called a **monoculture**.

One big disadvantage of monoculture is that the crop can be very easily harmed by pests and diseases. This means the farmer often has to use a lot of chemical pesticides to control them. Large fields and the use of pesticides can reduce the biodiversity of the environment, so is not sustainable in the long-term. Large-scale monoculture crop production could also threaten food supply if a natural disaster wiped out the whole crop.

Packaging – the problem

Packaging is useful: it protects food and other goods on their journey from farm or factory to warehouse or shop and then to our homes. But packaging can be an environmental problem. It is one area that could be made more sustainable. We need to think carefully about how we make things and how we dispose of the things we no longer need.

In 2008 an estimated 10.7 million tonnes of packaging was created in the UK. Plastics are light and strong so are used to package about 50% of all goods. All this packaging will end up as waste. In the past the majority of this waste ended up in landfill, a huge hole in the ground that is then covered over and landscaped, hiding the rubbish underground.

Packaging – improving sustainability

The use of **biodegradable** packaging materials such as paper and plant-based plastics, instead of oil-based plastics, would seem to be more sustainable. They should decompose if put in a landfill site, only releasing carbon dioxide that has recently been fixed in photosynthesis. So that it is not just a case of hiding the waste underground. But the bacteria that decompose waste need to have oxygen, and landfill sites often lack oxygen. This oxygen deficiency prevents the breakdown of the waste. Any decay that does occur in low-oxygen conditions produces methane, a powerful greenhouse gas.

One solution to the problem would be to recycle packaging waste. In 2008 the UK recycled 61% of its packaging waste, a massive increase from the 28% recycled in 1997.

It would be even better to reduce the amount of packaging used. This would reduce the amount of resources and energy used to make the original packaging materials and transport them. It would also avoid the need to collect, transport, and recycle the waste, saving more energy and reducing carbon dioxide emissions.

Biodegradable plastics are made from starch and cellulose from plants.

Newspapers dug up after years in a landfill site can still be read.

Questions

1 Explain what is meant by sustainability.

2 Give two reasons why it is important that we do not lose biodiversity of life on Earth.

3 Explain how the use of packaging can be made more sustainable.

Key words

- ✓ **biodiversity**
- ✓ **sustainability**
- ✓ **monoculture**
- ✓ **biodegradable**

Science Explanations

In this module you will consider different explanations for evolution and learn about natural selection. You will learn how living organisms are dependent on their environment and each other for survival, and you will learn about biodiversity and sustainability.

You should know:

- that a species is a group of breeding organisms, producing fertile offspring
- why all species within a food web are dependent on each other
- why organisms compete for resources with other species in the same habitat
- how organisms become extinct if they cannot adapt to environmental change, or if a competitor, predator, or disease-causing organism enters the environment
- that the Sun is the ultimate source of energy for nearly all organisms
- how energy transfers through the ecosystem when organisms are eaten or decay
- how energy is lost from a food chain as heat, waste products, and uneaten parts, limiting the length of the food chain
- how carbon cycles through the environment, including the processes of combustion, respiration, and photosynthesis
- that the nitrogen cycle involves nitrogen fixation, conversion to proteins, excretion, decay, uptake of nitrates by plants, and denitrification
- that life on Earth began 3500 million years ago and evolved from simple living things
- why all individuals are different and that some variation is genetic
- how mutations increase genetic variation and can be passed on to offspring
- how certain characteristics favour the survival of certain individuals in the process of natural selection
- how humans use selective breeding to choose characteristics in plants and animals
- how new species evolve through a combination of mutations, environmental changes, natural selection, and isolation
- that the analysis of DNA and the fossil record provides evidence for evolution
- how the classification of organisms shows their evolutionary relationship
- that Darwin's theory of evolution resulted from observations and creative thought
- that biodiversity includes the variation within and between different species
- how biodiversity ensures sustainability by increasing the stability of ecosystems and is vital for the development of food crops and medicines
- why large-scale monoculture of a single crop does not maintain biodiversity
- that all packaging materials use raw materials, energy for their production and transport, and create pollution; reducing them improves sustainability.

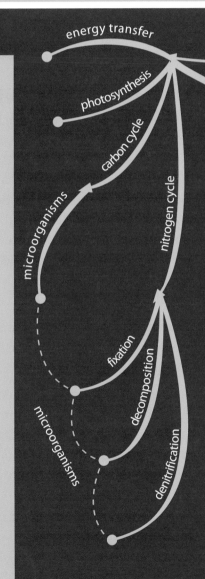

energy transfer
photosynthesis
carbon cycle
nitrogen cycle
microorganisms
microorganisms
fixation
decomposition
denitrification

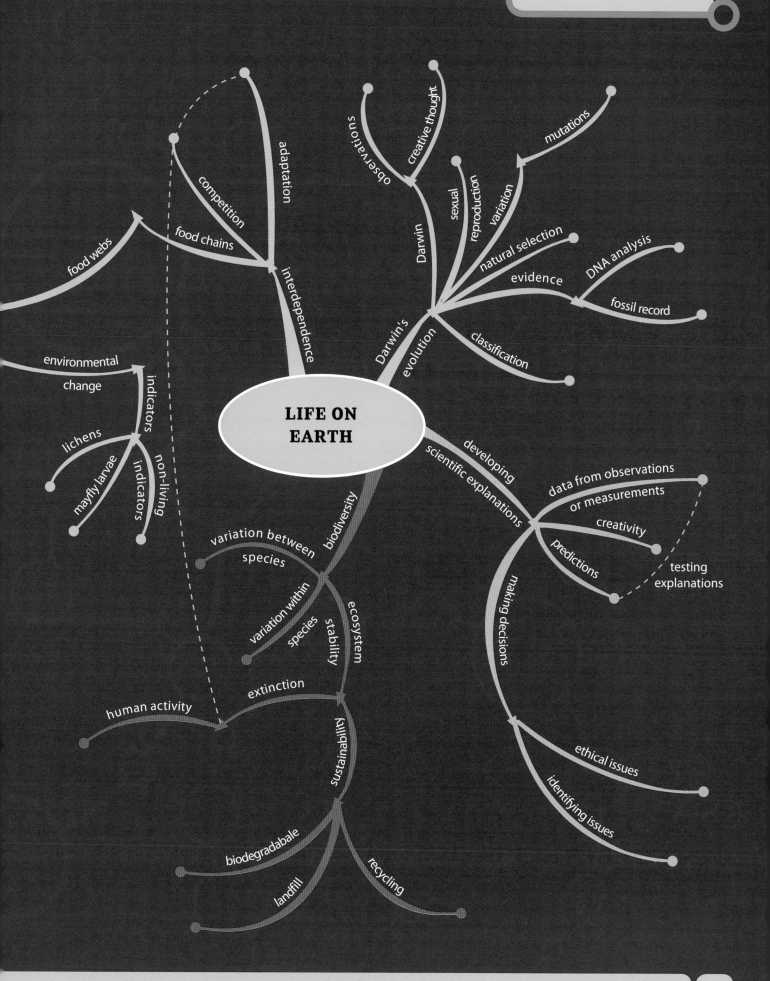

LIFE ON EARTH

adaptation

competition

food chains

interdependence

food webs

environmental change

indicators

lichens

mayfly larvae

non-living indicators

observations

creative thought

mutations

Darwin

sexual reproduction

variation

natural selection

evidence

DNA analysis

fossil record

Darwin's evolution

classification

developing scientific explanations

data from observations or measurements

creativity

predictions

testing explanations

making decisions

ethical issues

identifying issues

biodiversity

variation between species

variation within species

ecosystem stability

extinction

human activity

sustainability

biodegradabale

landfill

recycling

Ideas about Science

Science is about collecting data and using that data creatively to generate explanations. To test whether the explanations are correct, predictions are made and then new experiments or observations are checked against the predictions.

Scientists base their theories and explanations on observations and data. You will need to be able to distinguish between data and explanations and recognise those explanations that involve creative thinking.

Very often, conflicting explanations for the same data are produced. You should be able to suggest why scientists might disagree and to identify the better explanation, giving reasons for your choice. For example, Lamarck thought that characteristics were acquired during life and then passed on to their offspring. By thinking creatively, Darwin produced a new theory; he suggested that organisms evolved due to the process of natural selection. Some scientists disagreed with Darwin because of their personal or religious beliefs.

Scientific explanations can be tested by predictions and comparing the prediction with data obtained from experiments. When the data supports the prediction our confidence in the explanation is increased. For example, we could predict that when a new antibiotic is produced, bacteria will rapidly evolve resistance to the antibiotic by the process of natural selection. If this were to happen it would increase our confidence in the theory of evolution and the process of natural selection.

Science-based technologies improve the quality of our lives. However, science can sometimes have unintended and undesirable consequences. So the benefits of the new technology need to be weighed against the costs. Humans have introduced new animals into ecosystems with the best of intentions but the consequences have sometimes been terrible, such as the introduction of the rabbit and cane toad into Australia.

You will need to be able to suggest examples of unintended impacts of human activity on the environment, and use ideas and data about sustainability to compare the sustainability of different products or processes, for example, using biodegradable products in packaging.

Review Questions

1 The chart shows the evolution of humans (*Homo sapiens*) over the past 7 million years, drawn using evidence from fossils.

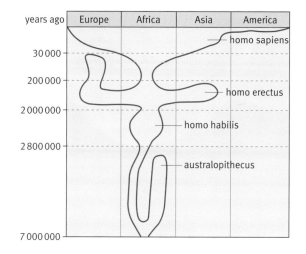

years ago	Europe	Africa	Asia	America
30 000				homo sapiens
200 000				
2 000 000				homo erectus
				homo habilis
2 800 000				
				australopithecus
7 000 000				

a Neanderthals are another extinct relative of humans. They did not evolve into *Homo sapiens*. Neanderthals became extinct just over 30 000 years ago. Identify the part of the chart that represents Neanderthals and state which continent they were mainly found in.

b Use the chart to answer these questions.

  i Which of the statements below are true?

    A All the species named on the chart evolved from a common ancestor.

    B *Homo sapiens* appeared before *Homo erectus*.

    C *Australopithicus* evolved from *Homo habilis*.

    D *Homo habilis* spread to more continents than *Homo sapiens*.

    E *Homo erectus* was mainly found in Africa.

  ii Name one species on the chart that is not yet extinct.

2 Look at the food web.

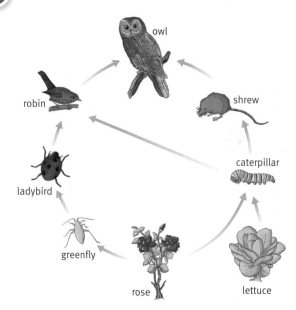

a Explain the effect on the food web of a farmer spraying the caterpillars with an insecticide.

b Write down one food chain found in the food web.

c Explain how energy enters the food chain, how it is transferred along the food chain, and how it is lost from it.

3 Explain the role of microorganisms in the nitrogen cycle.

You may draw a diagram to help your answer.

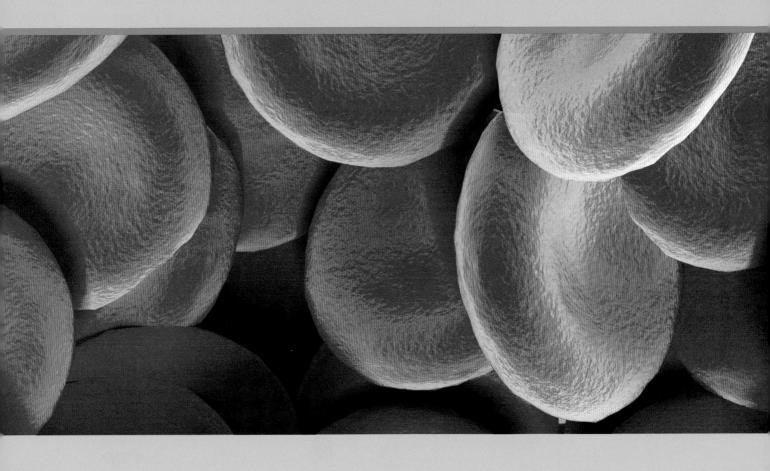

B4 The processes of life

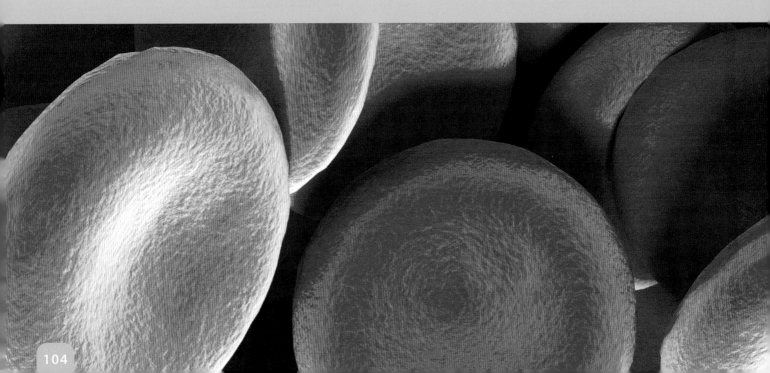

Why study the processes of life?

Look inside the cells of the largest animal or the smallest bacterium and you will find the same things happening. All living things need to be able to process molecules to generate energy, repair damage, and grow new cells. Enzymes speed up these chemical reactions. They need specific conditions to work at their best. Understanding the processes of life helps us to treat diseases and improve food production. It allows us to investigate the workings of all life on Earth.

What you already know

- Genes in the nuclei control how an organism's cells develop and function.

- Genes are instructions for a cell that describe how to make proteins.

- Living organisms are adapted to their environment and compete with one another.

- Nearly all organisms are ultimately dependent on energy from the Sun.

- Plants absorb a small percentage of the Sun's energy for the process of photosynthesis.

- Energy from photosynthesis is stored in the chemicals that make up the plants' cells.

Find out about

- the chemical processes that happen in all living cells

- the way that enzymes speed up chemical reactions

- how the process of respiration releases energy for cells to use

- how photosynthesis captures energy from the Sun and is the start of food chains

- how investigations can explore the relationship between changing variables and outcomes.

The Science

Photosynthesis powers all life on Earth. Respiration breaks down sugars and produces energy for cell processes. Photosynthesis captures energy from sunlight and uses it to combine carbon dioxide and water to make sugars. This energy is passed along the food chain.

Ideas about Science

To investigate the way outcomes are affected when certain factors are changed, it is important to control any other factors that may also affect the outcome.

Find out about

✓ **the processes carried out by all living things**

All living things are made up of **cells**. Some are just a single cell while others, like you, are made from billions of cells. You may not think you are like single-celled **bacteria** but there are some processes that all living things perform.

○ Move.

Bacteria and animals may move to find food, escape predators, or find better conditions to grow. Plants are rooted to the spot but grow to find sunlight.

○ Respire.

Energy is needed to carry out cell processes. Respiration is a series of chemical reactions that release energy from food molecules like sugar. Enzymes allow these reactions to happen in plant, animal, and microbial cells.

○ Sense.

Organisms need to be able to sense and respond to their surroundings. Plants grow towards sunlight and woodlice run away from it.

○ Grow.

Bacteria grow and divide to form new bacteria. Plants and animals are made from billions of cells. You have grown from a single fertilised egg cell.

○ Excrete.

Living cells produce wastes that are toxic if they build up. Wastes are removed by excretion. Carbon dioxide is a waste product of respiration.

○ Feed.

Living things need a supply of energy. They get this from their food. Plants make their own food during **photosynthesis**.

○ Reproduce.

All living things eventually die. Reproduction makes new generations that keep the species alive.

Life processes

There is a biological chemical factory in the cytoplasm of every plant cell. It takes small molecules and builds them into large molecules. It uses chemical reactions to build itself, copy itself, and repair itself. It takes large food molecules and breaks these down to release energy in respiration.

All living things are able to do this. Your own cells are performing an amazing set of chemical reactions and processes all the time. All these reactions in living organisms are catalysed by enzymes. Each cell has genes with instructions for making these enzymes.

Using life processes: making enzymes

Large vessels, like the fermenter in the picture below, are used to grow bacteria so that their enzymes can be harvested. They contain a nutrient solution, and conditions such as temperature, pH, and oxygen levels are controlled for optimum growth.

As the bacteria grow, they release their enzymes into the nutrient solution. When the nutrients are used up, the solution is filtered to remove the bacteria. The required enzymes can be purified from the solution. They are now available for use in applications such as the food industry, textiles industry, and in the home as biological washing powders.

In the cytoplasm of every one of this plant's cells is a biological chemical factory.

Bacteria are grown in this fermenter. They make enzymes that are used to make denim jeans look faded.

Like any living organism, bacteria need to take in nutrients. To digest their food, they produce and release enzymes into their surroundings. These break down any food molecules so that they can be absorbed into the microbe.

Key words
- ✓ cells
- ✓ bacteria
- ✓ respiration
- ✓ enzymes
- ✓ photosynthesis

Questions

1 List the processes that all living things carry out.

2 Which process generates energy for living cells?

3 Suggest ways in which you are able to sense and respond to your surroundings.

4 Suggest some ways that your life would be different if humans had not developed a good understanding of the life processes that happen inside all cells.

Find out about

- ✓ why you cannot live without enzymes
- ✓ how enzymes work

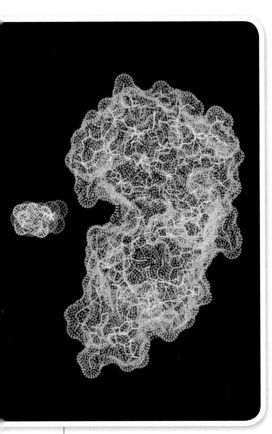

A computer graphic of an enzyme, its active site, and the product of a reaction.

The lock-and-key model of enzyme function. (*Note:* This diagram is schematic. This means that if you were able to see the molecules and enzymes, they would not look like they do here. Enzymes consist of many molecules – see the graphic above left for an idea of scale. However, this is a useful way of picturing, or modelling, how an enzyme works.)

The chemical reactions that take place in cells rely on **enzymes**. Every cell has genes with instructions for making enzymes. Enzymes work best in certain conditions, for example, at a certain temperature. This is an important reason why you need a steady state inside your cells.

What are enzymes?

Enzymes are the **catalysts** that speed up chemical reactions in living organisms. They are **proteins** – large molecules made up of long chains of **amino acids**. The amino acid chains are different in each protein, so they fold up into different shapes. An enzyme's shape is very important to how it works. The sequence of amino acids in each enzyme is determined by instructions in a gene.

How do enzymes work?

Some enzymes break down large molecules into smaller ones. Others join small molecules together. In all cases, the molecules must fit exactly into a part of the enzyme called the **active site**. It is a bit like fitting the right key in a lock. So scientists call the explanation of how an enzyme works the **lock-and-key model**.

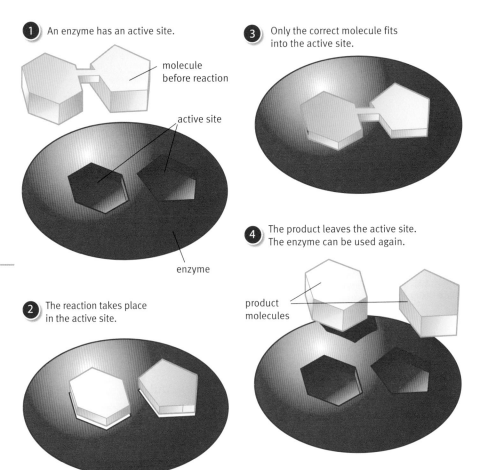

1 An enzyme has an active site.

molecule before reaction

active site

enzyme

2 The reaction takes place in the active site.

3 Only the correct molecule fits into the active site.

4 The product leaves the active site. The enzyme can be used again.

product molecules

Why do we need enzymes?

At 37 °C, chemical reactions in your body would happen too slowly to keep you alive.

One way of speeding up a reaction is to increase the temperature. A higher body temperature could speed up the chemical reactions in your body. But higher temperatures damage human cells. Also, to keep your body warm you have to release energy from **respiration**. For a higher body temperature you would need a lot more food to fuel respiration.

So, we rely on enzymes to give us the rates of reaction that we need. They can increase rates of reaction by up to 10 000 000 000 times. It is not possible to live without enzymes.

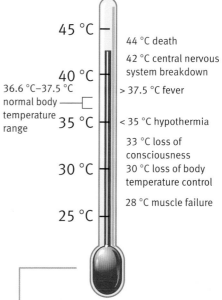

45 °C
44 °C death
42 °C central nervous system breakdown
40 °C
> 37.5 °C fever
36.6 °C–37.5 °C normal body temperature range
35 °C
< 35 °C hypothermia
33 °C loss of consciousness
30 °C
30 °C loss of body temperature control
28 °C muscle failure
25 °C

Your core body temperature is about 37 °C, but a small variation either side of this is normal. A core temperature over 42 °C or under 28 °C usually results in death.

Around 80% of the energy from your food is used for keeping warm. If you had to maintain a higher body temperature, you would have to spend a lot more time eating to provide the calories to heat your body.

Shrews have a large surface area for their volume. They lose heat to the environment over their whole body surface. To release enough energy to maintain their body temperature, they have to eat 75% of their body mass in food each day. If not, they die within two to three hours.

Questions

1 Write down:
  a what enzymes are made of
  b what enzymes do.

2 Explain how an enzyme works. Use the key words on this page in your answer.

3 The enzyme amylase breaks down **starch** to sugar (maltose); catalase breaks down hydrogen peroxide to water and oxygen. Explain why catalase does not break down starch.

Key words
- ✔ **proteins**
- ✔ **amino acids**
- ✔ **active site**
- ✔ **lock-and-key model**
- ✔ **starch**
- ✔ **catalysts**

Find out about

- ✓ **how temperature and pH affect enzymes**

'Ice fish' such as Antarctic cod are active at 2 °C because their enzymes work best at low temperatures. Most organisms would be dead or very sluggish at 2 °C.

Bacteria living in hot springs have enzymes that withstand high temperatures.

How does temperature affect enzymes?

At low temperatures, enzyme reactions get faster if the temperature is increased.

But above a certain temperature the reaction stops. This is because enzymes are proteins. Higher temperatures change an enzyme's shape so that it no longer works. The diagram below explains what happens using the lock-and-key model.

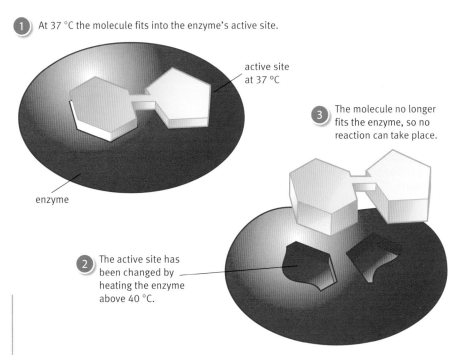

1 At 37 °C the molecule fits into the enzyme's active site.

active site at 37 °C

3 The molecule no longer fits the enzyme, so no reaction can take place.

enzyme

2 The active site has been changed by heating the enzyme above 40 °C.

How an enzyme reaction can be stopped by a rise in temperature.

Enzymes are denatured at high temperatures

High temperatures change the shape of an enzyme. They do not destroy it completely. But even when an enzyme cools down, it does not go back to its original shape. Like cooked egg white, the protein cannot be changed back. The enzyme is said to be **denatured**.

Why 37 °C?

The temperature at which an enzyme works best is called its **optimum temperature**. Enzymes in humans work best at around 37 °C. Some organisms have cells and enzymes adapted to different temperatures.

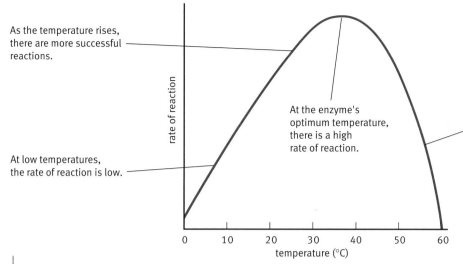

As the temperature rises, there are more successful reactions.

At the enzyme's optimum temperature, there is a high rate of reaction.

At high temperatures the enzyme's active site is changed, so the molecules it catalyses cannot fit into the active site. The rate of reaction falls to zero as the enzyme is denatured by the high temperature.

At low temperatures, the rate of reaction is low.

The optimum temperature of an enzyme. The reaction rate is plotted at different temperatures – all other conditions are kept constant.

pH also affects enzymes

Proteins including enzymes can be damaged by acids and alkalis. The shape of an enzyme's active site will be changed if bonds holding the protein chains together are broken. The substrate will no longer be able to fit, so the enzyme becomes denatured. Every enzyme has an optimum pH at which it works best.

Enzyme	What it does	Optimum pH
salivary amylase	breaks down starch to sugar (maltose)	4.8
pepsin	breaks down proteins into short chains of amino acids	2.0
catalase	breaks down hydrogen peroxide into water and oxygen	7.6

Key words
- ✔ denatured
- ✔ optimum temperature

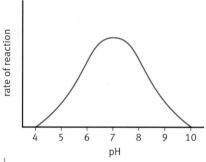

The enzyme used in this reaction has an optimum pH of 7.0.

Questions

1 Explain why increasing the temperature causes enzyme reactions to stop at higher temperatures.

2 What is meant by an enzyme's optimum temperature?

3 Enzymes in food and enzymes from microorganisms such as bacteria and fungi make food decay. Why does food stay fresh in a refrigerator for a few days but for months in a freezer?

4 Starch synthase is an enzyme found in potato tubers, which grow in the UK. The enzyme joins glucose molecules together to store them as starch. Some students compared its rate of reaction at 5 °C, 20 °C, and 45 °C.

Think about the environment that the potato grows in and suggest what would happen if the students:

a raised the temperature of the tuber at 5 °C to 20 °C?

b raised the temperature of the tuber at 20 °C to 45 °C?

Every second of every day, enzymes catalyse billions of chemical reactions inside every cell of your body. Plants are just as active too. They have some different genes with instructions for making special enzymes. Without the types of reactions that go on inside plant cells, humans and other animals would not exist.

Photosynthesis

Plants capture energy from sunlight. This may look as easy as sunbathing, but the process is one of nature's cleverest tricks. Many complex chemical reactions are involved.

During photosynthesis, plants trap energy from sunlight and use it to make all the molecules they need for growth. These include sugars, starch, enzymes, and **chlorophyll**. These molecules feed other organisms along the food chain. So photosynthesis supplies food for life on Earth.

What happens during photosynthesis?

The chemical equation for photosynthesis is:

$$6CO_2 + 6H_2O \xrightarrow[\text{chlorophyll}]{\text{light energy}} C_6H_{12}O_6 + 6O_2$$

carbon dioxide   water   glucose   oxygen

Do not let the equation mislead you. The reaction does not happen in one go – it has lots of smaller steps. The equation is a convenient way of summing up the process.

A **glucose** molecule is made up of carbon, hydrogen, and oxygen atoms. So glucose is a carbohydrate.

Photosynthesis takes place in **chloroplasts**. They contain a green pigment called chlorophyll. Chlorophyll absorbs light and uses the energy to kick-start photosynthesis.

Energy from light splits water molecules into hydrogen and oxygen atoms. The hydrogen is combined with carbon dioxide from the air to make glucose. The oxygen is released as a waste product. It passes out of the plant into the air. Given enough raw materials, light, and the right temperature, a large tree can make 2000 kg of glucose in a day.

Find out about

- ✔ how photosynthesis captures energy for life on Earth
- ✔ what happens to the glucose made by photosynthesis

Key words

- ✔ chlorophyll
- ✔ chloroplasts
- ✔ cellulose
- ✔ vacuole
- ✔ cell wall
- ✔ cell membrane
- ✔ nucleus
- ✔ cytoplasm
- ✔ glucose

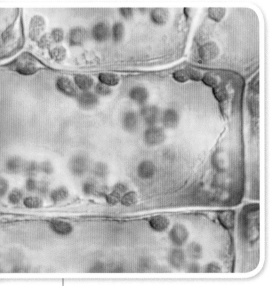

Chloroplasts contain the green pigment chlorophyll and the enzymes that are needed for photosynthesis. (Magnification × 2000.)

Using glucose from photosynthesis

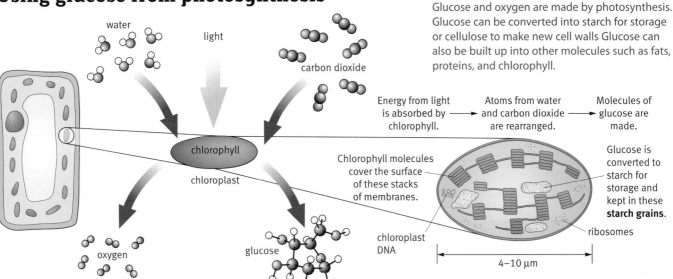

Glucose and oxygen are made by photosynthesis. Glucose can be converted into starch for storage or cellulose to make new cell walls Glucose can also be built up into other molecules such as fats, proteins, and chlorophyll.

Energy from light is absorbed by chlorophyll. → Atoms from water and carbon dioxide are rearranged. → Molecules of glucose are made.

Chlorophyll molecules cover the surface of these stacks of membranes.

Glucose is converted to starch for storage and kept in these **starch grains**.

chloroplast DNA

ribosomes

4–10 μm

Glucose made during photosynthesis is used by plant cells in three ways.

(1) Making other chemicals needed for cell growth

Glucose is converted into other carbohydrates, as well as fats and proteins. Two important carbohydrates in plants are **cellulose** and starch. Cellulose and starch are both polymers of glucose. They are made up of thousands of glucose molecules linked together.

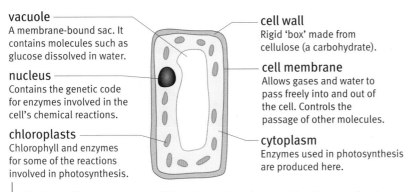

vacuole
A membrane-bound sac. It contains molecules such as glucose dissolved in water.

nucleus
Contains the genetic code for enzymes involved in the cell's chemical reactions.

chloroplasts
Chlorophyll and enzymes for some of the reactions involved in photosynthesis.

cell wall
Rigid 'box' made from cellulose (a carbohydrate).

cell membrane
Allows gases and water to pass freely into and out of the cell. Controls the passage of other molecules.

cytoplasm
Enzymes used in photosynthesis are produced here.

A plant cell contains many different structures involved in photosynthesis.

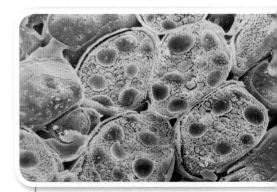

Starch grains in a plant cell store glucose as starch. (Magnification × 200.)

(2) Storing energy in starch molecules

Any excess glucose is converted into starch. Starch is a storage molecule. Starch can be converted back to glucose when needed. Starch is stored in leaf cells, but some plants have special organs, such as the tubers of a potato, which have cells that are filled with starch.

(3) Releasing energy in respiration

Glucose molecules are broken down by respiration, releasing the energy stored in the molecules. This energy is used to power chemical reactions in the cells, such as converting glucose to cellulose, starch, or proteins.

Questions

1 Write down the word equation that sums up photosynthesis.

2 Draw a diagram to show the flow of chemicals in and out of leaves during photosynthesis.

3 Describe the main stages in making glucose by photosynthesis.

4 Glucose from photosynthesis has three roles in the plant cell. Explain what these are.

Find out about

- ✓ **how chemicals move in and out of cells**
- ✓ **how carbon dioxide and oxygen move in and out of a leaf during photosynthesis**

① Water just poured onto tea bag

To start with, dissolved molecules from the tea are concentrated close to the tea bag. There are few tea molecules in the rest of the hot water.

② About 30 seconds later

The tea molecules move from where they are highly concentrated into regions where they are less concentrated.

③ About 2 minutes later

After a few minutes the tea molecules spread evenly throughout the cup.

Molecules move in and out of cells all the time. Cells need a constant supply of raw materials for chemical reactions, and waste needs to be removed.

Photosynthesis in a plant leaf cell uses carbon dioxide and produces waste oxygen. Movement of these molecules takes place by the process of **diffusion**.

Diffusion

Molecules in gases and liquids move about randomly. They collide with each other and change direction. This makes them spread out.

Overall, more molecules move away from where they are concentrated than move the other way. The molecules diffuse from areas of their high concentration to areas of low concentration.

For example, molecules diffuse out of a tea bag when you make a cup of tea. Diffusion is a passive process. It does not need any extra energy.

This swimmer's cells need oxygen and glucose for respiration. They must get rid of carbon dioxide. These molecules move in and out of cells by diffusion.

Supplying photosynthesis

Photosynthesis is a vital process that takes place in leaves using the energy from sunlight. Special structures in leaves allow the chemicals needed for photosynthesis to be delivered and waste oxygen to be removed by diffusion.

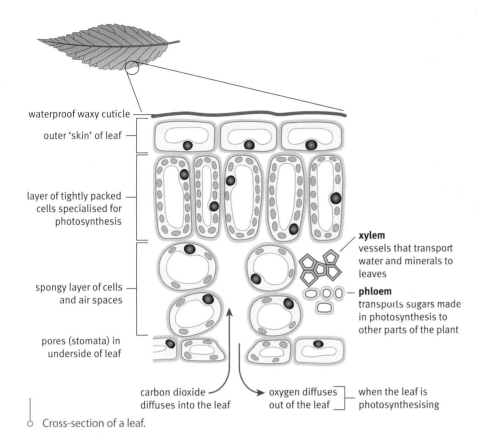

- waterproof waxy cuticle
- outer 'skin' of leaf
- layer of tightly packed cells specialised for photosynthesis

xylem
vessels that transport water and minerals to leaves

phloem
transports sugars made in photosynthesis to other parts of the plant

- spongy layer of cells and air spaces
- pores (stomata) in underside of leaf

carbon dioxide diffuses into the leaf

oxygen diffuses out of the leaf

when the leaf is photosynthesising

Cross-section of a leaf.

Photosynthesis, diffusion, and gas exchange in leaves

The underside of a leaf contains thousands of tiny holes called **stomata**. These allow carbon dioxide into the leaf and oxygen out. Diffusion drives these gases from high to low concentrations.

	Carbon dioxide	Oxygen
Cells in leaf during photosynthesis	• Carbon dioxide is used in cells for photosynthesis. • Low concentration of carbon dioxide in cells in the leaf. Carbon dioxide diffuses through the stomata and into cells in the leaf to supply photosynthesis. *into leaf*	• Oxygen is produced during photosynthesis. • High concentration of oxygen builds up in cells of the leaf. Oxygen diffuses out of the cells and through the stomata. *out of leaf*
Air surrounding the leaf	• Higher concentration of carbon dioxide in the air.	• Lower concentration of oxygen in the air.

Key word
✓ **diffusion**

Questions

1 Write down a definition for diffusion.

2 Name three chemicals that move in and out of cells by diffusion.

3 Explain how diffusion lets you smell the vinegar from fish and chips on a plate in front of you.

4 Earthworms do not have lungs. They rely on diffusion through their skin to exchange oxygen and carbon dioxide with air spaces in the soil. Use a labelled diagram to explain how diffusion takes oxygen into an earthworm and carbon dioxide out.

Water is a vital component of all living things. Without water, the chemical reactions that take place inside cells could not happen.

Osmosis is a specific type of diffusion. It is the process that moves water molecules into and out of cells. The cell membrane is important in osmosis.

Cell membranes are partially permeable

Cell membranes let some molecules through but block others. Tiny channels in the membrane allow small molecules, like water, to travel through them. Larger molecules are too big and cannot get through. Cell membranes are **partially permeable membranes**.

Diagrams **a** and **b** below show how water molecules move during osmosis. In this example the membrane allows water through but the glucose molecules are too big to get through.

Key

partially permeable membrane allows some molecules through and acts as a barrier to others

glucose molecule

○ water molecule

 water molecules associated with glucose molecule (these molecules are not free to move by osmosis)

(*Note:* In these diagrams, the circles represent molecules, not individual atoms. Cell membranes are also made of molecules.)

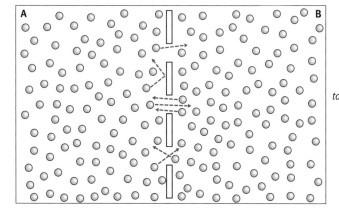

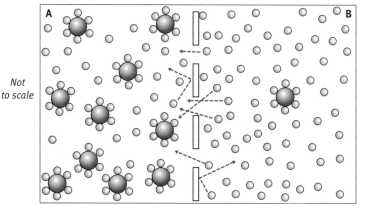

Not to scale

a This membrane is separating water molecules. Water molecules move at random. As many pass from left (A) to right (B) as pass from right (B) to left (A).

b This membrane is separating two glucose solutions. There are more free water molecules and fewer glucose molecules on the right (B). In other words, water molecules that are free to move by osmosis are in higher concentration on the right. So there is overall movement of water from right (B) to left (A).

Movement of water molecules

A solution with a high concentration of water molecules that are free to move is a **dilute** solution. For example, if you make a dilute drink of squash, it has lots of water in it. A more **concentrated solution** of squash would have fewer free water molecules.

More water molecules move away from an area of higher concentration of free water molecules. Think of it as diffusion of water. This overall flow of water from a dilute to a more concentrated solution across a partially permeable membrane is called osmosis.

Osmosis in plant cells

Plants do not have skeletons to give them support. They must keep their structure by having cells that are just the right size and shape. Osmosis is important in this. It drives the uptake of water by plant roots and determines how water passes from one cell to another throughout the whole plant.

If plant cells take in too much water, they bulge and become stretched. Their strong cell wall prevents them from bursting. If plant cells lose too much water, they shrink. Plant cells need to keep just the right amount of water inside their cytoplasm.

Look at the two pictures on the right to see what happens to a plant if it does not have enough water.

Glucose is stored as starch

Plants make glucose during photosynthesis. The glucose is transported from the leaves to other cells where it is stored until it is needed for respiration. This poses a problem for the plant cells that store the glucose. They will take in too much water by osmosis. The water would move from a dilute solution surrounding the cells into the more concentrated glucose solution in the cells. To overcome this problem, glucose is stored as starch.

Large carbohydrates like starch are **insoluble**. They have very little effect on the concentration of the solutions in a plant cell. This makes them ideal for storing glucose. The starch does not affect the movement of water in and out of the cells. Starch is kept in small, membrane-bound bags in the cell called **starch grains**.

The photographs show the same plant. In the lower image, the plant is shown after it has not been watered for 10 days. Notice how the structure has changed as the plant cells have dried.

Questions

1 Explain what is meant by a partially permeable membrane.

2 Write down a definition of osmosis.

3 A student put a raisin (a dried grape) into a glass of water. They noticed that the raisin expanded and swelled. Explain this observation. Include a diagram to show the movement of water.

4 Explain why starch is needed to store glucose in plant cells.

5 Water moves into root cells by osmosis. What does this tell you about the water in the soil?

Find out about

- why plants need minerals
- how minerals are absorbed in the roots by active transport

Nitrates contain this group of atoms:

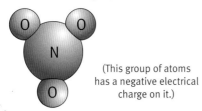

(This group of atoms has a negative electrical charge on it.)

Nitrate ions are found dissolved in soil water, and in rivers and seas.

Plants capture energy from sunlight during photosynthesis. This energy builds glucose for respiration. Glucose also supplies the raw materials needed to make other molecules like proteins, fats, and DNA. These molecules need elements contained in minerals from the soil.

Making proteins needs nitrogen

Proteins are long chains of amino acids. To make amino acids, nitrogen must be combined with carbon, hydrogen, and oxygen atoms from glucose made during photosynthesis.

Most of the Earth's nitrogen is in the air, but plants mainly take in nitrogen from the soil as **nitrate ions**. These nitrates are absorbed by **root hair cells**.

Nitrates are not the only minerals that plants need. For example, they need magnesium to make chlorophyll and phosphates to make DNA. As proteins are used to build cells and make enzymes, nitrates are needed in the largest quantities. Fertilisers contain minerals such as phosphates and nitrates.

Nitrate ions are absorbed by active transport

Plant roots absorb nitrate ions dissolved in water in the soil. Nitrate ions are at a higher concentration inside the root cells than in the surrounding soil. Diffusion would move the nitrate ions out of the roots and into the soil. To overcome this, the cells use a process called **active transport** to pump nitrates from the soil and into the roots.

magnified section of a plant root showing root hair cells

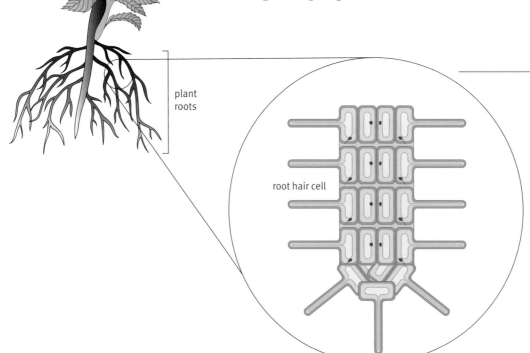

plant roots

root hair cell

Plant roots have a very large surface area to help them absorb water and minerals. Tiny root hair cells take in minerals, such as nitrates, by active transport.

Another way of getting molecules into cells

When a cell needs to take in molecules that are in higher concentration inside the cell than outside, active transport is used.

In active transport, cells use the energy from respiration to transport molecules across the membrane. An example of active transport is where nitrates are taken into plant roots against their diffusion gradient.

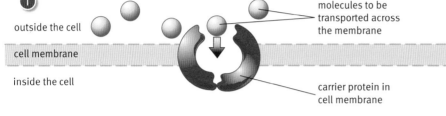

outside the cell

molecules to be transported across the membrane

cell membrane

inside the cell

carrier protein in cell membrane

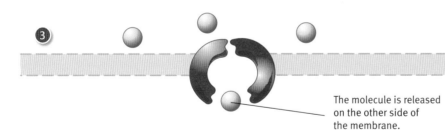

ENERGY

Energy from respiration is used to change the shape of the carrier protein.

The molecule is released on the other side of the membrane.

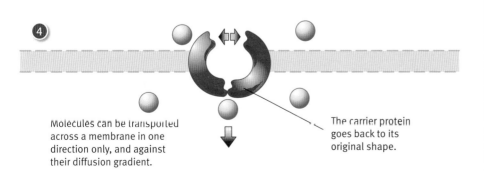

Molecules can be transported across a membrane in one direction only, and against their diffusion gradient.

The carrier protein goes back to its original shape.

Not to scale

Movement of molecules across a cell membrane by active transport (schematic).

Questions

1. Explain why cells sometimes need to use active transport.

2. Name one chemical that is moved into cells by active transport.

3. Write down two main differences between diffusion and active transport.

4. Why do plant cells need a source of nitrate ions?

Find out about

✓ **what limits the rate of photosynthesis**

Intensive tomato farming takes place all year round in this greenhouse.

The conditions inside the greenhouse in the photograph on the left are kept under very careful control. The tomato plants growing here have the optimum conditions for photosynthesis. They are making glucose at their highest rate, so they are growing quickly. All this is planned by the farmer so that the yield from the tomato plants will be as high as possible. Yield is the amount of product the farmer has to sell.

All reactions speed up when the temperature rises, and photosynthesis is no exception. The greenhouse is kept warm at 26 °C. This is the optimum temperature for photosynthesis to take place in these plants. Some plants don't grow in the UK because the temperature is too cold. They are better adapted for life in hot climates. The temperature falls as you climb a mountain. Above a certain height it is too cold for many large plants to photosynthesise effectively.

Faster photosynthesis – light intensity

Other factors have an effect on the rate of photosynthesis. Energy from light drives photosynthesis, so increasing the amount of light a plant receives increases the rate of photosynthesis.

The diagram below shows an experiment to investigate how changing **light intensity** affects the **rate of photosynthesis** in a piece of pondweed. The results from the experiment are shown in the graph. The graph shows that:

- at low light intensities, increasing the amount of light increases the rate of photosynthesis
- at a certain point increasing the amount of light stops having an effect on the rate of photosynthesis.

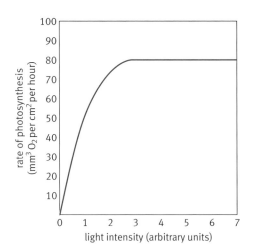

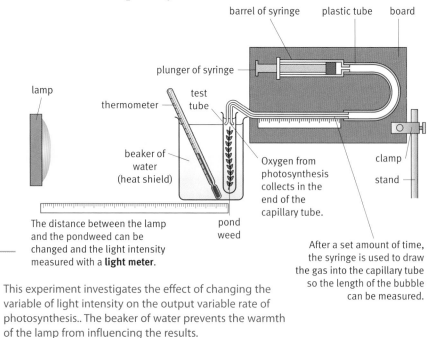

The distance between the lamp and the pondweed can be changed and the light intensity measured with a **light meter**.

This experiment investigates the effect of changing the variable of light intensity on the output variable rate of photosynthesis.. The beaker of water prevents the warmth of the lamp from influencing the results.

Why does the rate not keep on rising?

Photosynthesis needs more than energy from light. Extra light makes no difference to the rate of photosynthesis if the plant does not have the carbon dioxide, water, or chlorophyll to use the energy from light to the full. The temperature must also be high enough for photosynthesis reactions to speed up. Increasing the light intensity stops having an effect on the rate of photosynthesis because one of these other factors is in short supply. This factor is called the **limiting factor**.

Limiting factors

In a British summer the limiting factor for photosynthesis is often water. Stomata close to prevent water diffusing out of the leaves, and this also has the unavoidable consequence of reducing carbon dioxide diffusion into the leaf.

The graph below shows the effect of increasing light intensity on the rate of photosynthesis at two different carbon dioxide concentrations. At 0.04% CO_2, more light increases the rate of photosynthesis up to a point, until light is no longer the limiting factor. Increasing the CO_2 level to 0.4% makes the rate of photosynthesis higher – CO_2 must have been the limiting factor. But even this graph levels off as another factor becomes in short supply.

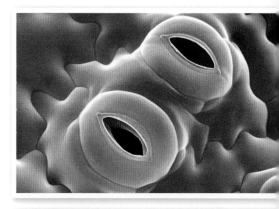

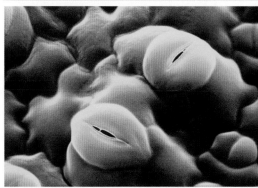

Top: The stomata on the underside of leaves open to allow gases to move in and out of the leaf.
Bottom: They close to conserve water. (Magnification × 400.)

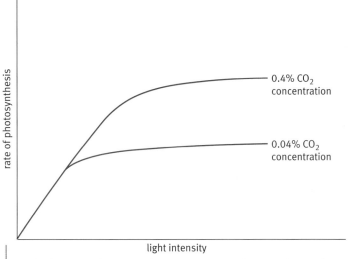

At the higher carbon dioxide concentration, photosynthesis takes place faster. But the rate still levels off. Another factor must be limiting photosynthesis.

(Graph axes: y-axis "rate of photosynthesis", x-axis "light intensity"; curves labelled "0.4% CO_2 concentration" and "0.04% CO_2 concentration".)

Carbon dioxide levels in the greenhouse

Carbon dioxide forms 0.04% of normal air. Levels over about 1% are toxic to plants and animals. The levels in the tomato greenhouse are kept at 0.1%. Raising the concentration higher than this has no effect on the rate of photosynthesis. So it would not be cost effective for the farmer to add more carbon dioxide than this to the greenhouse.

Key words
- ✓ light intensity
- ✓ rate of photosynthesis
- ✓ light meter
- ✓ limiting factor

Questions

1. Write down four factors that can affect the rate of photosynthesis.

2. Explain what is meant by a limiting factor.

3. Suggest a factor that could be limiting bluebells growing on a woodland floor in spring.

Find out about

- ✔ **how environmental conditions affect the plants that are able to grow**
- ✔ **ways to survey plants in a location**

Why do different plants grow in different locations?

Don't plants just grow where we plant them? Well that is the case in gardens and parks, but not in natural ecosystems. The great variety of plants has evolved to take advantage of **habitats** all over the Earth.

Some plants grow well in shade and others need bright light. Some plants need plenty of water and others can survive in deserts. All plants need minerals, water, and light in just the right amounts to be able to grow well.

The shading effect of trees determines the types of plant that can grow and survive in woodland.

Different habitats, different conditions

Woodland is a rich habitat for many plants and animals. Sunlight gets absorbed by tall trees, which hold their leaves up high for maximum photosynthesis. Breaks in the trees allow sunlight to pass through, but most of the woodland floor is in shade. Only specialised plants are able to grow and survive in these conditions.

Investigating different habitats

To understand why plants grow in particular locations, factors like soil pH, temperature, light intensity, and the availability of water are measured. Enough individual **samples** must be taken to get a true picture of what the conditions are like.

Key words

- ✔ habitat
- ✔ samples
- ✔ quadrat
- ✔ random
- ✔ transect

A square grid, called a **quadrat**, is used to survey the plants in a square metre. The quadrat is placed on the ground. Plants and animals within the quadrat are identified accurately and counted. An identification key can help to name the organisms. The key has descriptions or pictures that can be compared with the specimen. Plant growth is often recorded as percentage (%) cover.

The positioning of the quadrat in the area being investigated is **random**. This allows reliable comparisons between different locations. Placing quadrats randomly removes bias. Recording the plants and animals in a quadrat involves accurate identification of each species.

This scientist is using a quadrat to record the plants growing in a square metre.

Sometimes samples are taken at regular intervals along a straight line called a **transect**. This is useful when looking at how the types of plants change gradually from one area to another, for example, when moving from the shaded part of a wood into an open field. A lightmeter could be used to accurately measure light intensity. The sensor should be held at the same angle and the readings taken one after another so as to compare like with like.

Comparing bluebell growth

The results of a survey of bluebell growth are presented in the table below. Two areas were compared – one in shady woodland and the other in the middle of a field.

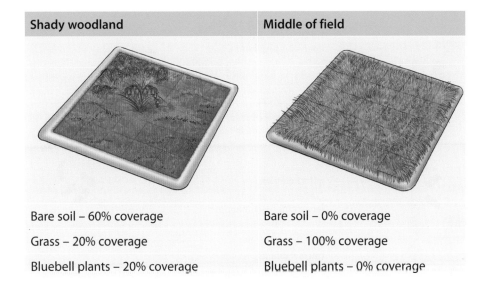

Shady woodland	Middle of field
Bare soil – 60% coverage	Bare soil – 0% coverage
Grass – 20% coverage	Grass – 100% coverage
Bluebell plants – 20% coverage	Bluebell plants – 0% coverage

The findings show that bluebells can survive in the shade. In the field, the bluebells cannot compete with the grass.

Questions

1 What do plants need to be able to grow?

2 What are the different environmental conditions that could affect plant growth?

3 Suggest how you could make sure that plant and animal data recorded at a location gives a reliable indication of the conditions in the area.

4 Cactus plants have thick waxy skins to prevent water loss. They can be grown indoors in the UK. Suggest what would happen if a cactus was planted in a flower bed outside.

Find out about

- ✓ **aerobic respiration in plant and animal cells**

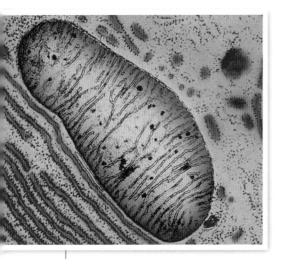

Some of the reactions for respiration take place in the cell cytoplasm, and many happen inside **mitochondria**. This electron micrograph shows a single mitochondrion. (Magnification × 64000.).

Photosynthesis combines carbon dioxide and water to make glucose. This captures the energy held in sunlight and converts it into energy held in glucose. Cells release the energy in glucose in a process called respiration. This is a carefully coordinated series of chemical reactions that happen in the cells of all living organisms.

Respiration releases energy in a form that cells can use for processes like active transport, movement, and building molecules used for growth and repair.

Aerobic respiration

Your body is a demanding animal. Billions of cells each carry out thousands of chemical reactions every second to keep you alive. Your cells need a constant supply of energy to drive these reactions. The food you eat provides you with molecules to make new cells. Food is also a store of chemical energy. This is converted by respiration into energy your cells can use.

Most of your energy comes from **aerobic respiration**. During aerobic respiration, glucose from food reacts with oxygen. The reactions release energy from the glucose. Respiration can be summarised by the following equation:

$$C_6H_{12}O_6 + 6O_2 \longrightarrow 6CO_2 + 6H_2O \text{ (+ energy released)}$$

glucose   oxygen   carbon dioxide   water

Respiration is a long series of reactions. It is summarised by this equation.

cytoplasm
Where enzymes are made. Location of reactions in anaerobic respiration.

cell wall

100 µm

chloroplast

nucleus
Holds genetic code for enzymes involved in respiration.

mitochondrion
Contains the enzymes for aerobic respiration.

cell membrane
Allows gases and water to pass freely into and out of the cell. Controls the passage of other molecules.

Structures used in respiration in plant and animal cells.

Typical plant cell

Typical animal cell

What happens to the energy from respiration?

All respiration releases energy from glucose. This energy will be needed by many processes in the cell.

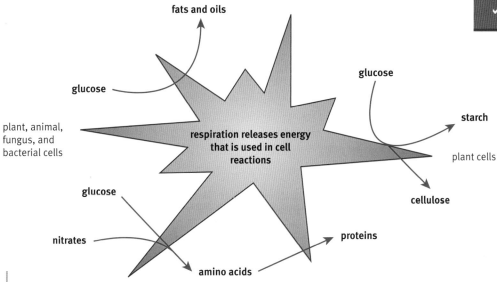

plant, animal, fungus, and bacterial cells

glucose → fats and oils

respiration releases energy that is used in cell reactions

glucose → starch — plant cells

glucose → cellulose

glucose, nitrates → amino acids → proteins

The energy from respiration is used in reactions that make many different molecules. These include **polymers** of glucose (starch and cellulose) and proteins, fats, and oils.

This girl is using energy for many different actions – such as processing information in her brain, moving, maintaining a constant internal environment, and growing and repairing her tissues.

Questions

1 Write down three processes for which your body needs energy.

2 a Write down the word equation for aerobic respiration.
  b Annotate the equation to show where the reactants come from, and what happens to the products.

3 Explain why the uptake of nitrate ions in a plant's roots by active transport requires energy, but the movement of carbon dioxide into leaf cells when they are photosynthesising does not.

Find out about

✓ **anaerobic respiration in humans and other organisms**

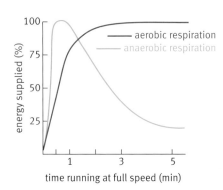

energy supplied (%)

— aerobic respiration
— anaerobic respiration

time running at full speed (min)

Germinating seeds respire anaerobically.

Many animals use a different type of respiration for short bursts of intense energy, for example, when a predator runs after its prey and the prey runs for its life. In these situations the animals' muscles cannot get oxygen quickly enough for aerobic respiration. So they switch to a form of respiration that does not need oxygen. It is called **anaerobic respiration**.

glucose $\longrightarrow$ lactic acid (+ energy released)

Anaerobic respiration in animals is summarised by this equation.

In animals, anaerobic respiration can only be used for a short period of time. It releases much less energy from each gram of glucose than aerobic respiration. Also, the waste product **lactic acid** is toxic in large amounts. If it builds up in muscles, it makes them feel tired and sore.

Anaerobic respiration in plants and microorganisms

Other organisms can also use anaerobic respiration when there is limited oxygen available, for example:

*   parts of plants, such as roots in waterlogged soils, and germinating seeds
*   some microorganisms, such as **yeast** (brewing and baking), lactobacilli (cheese- and yoghurt-making), and bacteria in puncture wounds.

Anaerobic respiration in plants, yeast, and some bacteria produces **ethanol** and carbon dioxide instead of lactic acid.

glucose $\longrightarrow$ ethanol + carbon dioxide (+ energy released)

Anaerobic respiration in plants and microorganisms is summarised by this equation.

The 100-m sprint takes about 10–11 seconds. The heart and lungs cannot increase oxygen supply to the muscles fast enough, so most of the energy required during the short race comes from anaerobic respiration.

Bacteria and yeast

Bacteria and yeast have been used for thousands of years in the preparation of foods such as bread, cheese, yoghurt, alcoholic drinks, and vinegar. Many of these foods are products of anaerobic respiration.

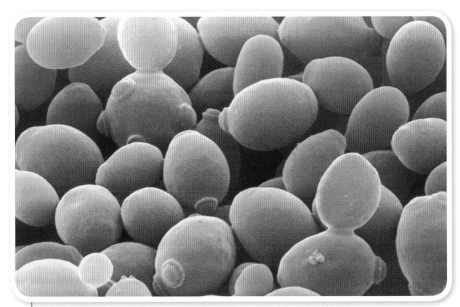

Yeast cells can respire anaerobically until ethanol builds up and becomes too toxic.

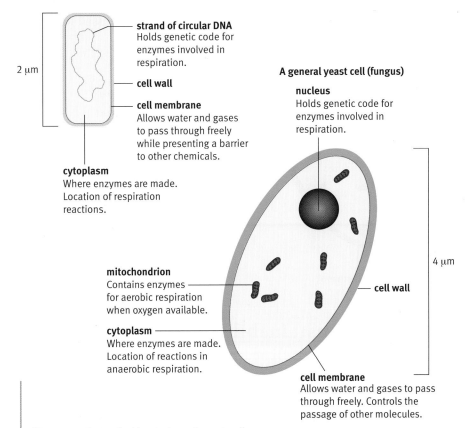

A general bacterial cell

2 µm

strand of circular DNA
Holds genetic code for enzymes involved in respiration.

cell wall

cell membrane
Allows water and gases to pass through freely while presenting a barrier to other chemicals.

cytoplasm
Where enzymes are made. Location of respiration reactions.

A general yeast cell (fungus)

nucleus
Holds genetic code for enzymes involved in respiration.

mitochondrion
Contains enzymes for aerobic respiration when oxygen available.

cytoplasm
Where enzymes are made. Location of reactions in anaerobic respiration.

cell wall

4 µm

cell membrane
Allows water and gases to pass through freely. Controls the passage of other molecules.

Structure of a typical bacteria and yeast cell.

Questions

1 Describe conditions where anaerobic respiration is an advantage to:
   a human beings
   b another organism.

2 Give an example of anaerobic respiration that people draw on to make a useful product.

3 Yeast is used in brewing to make alcoholic drinks. Suggest why the yeast stops growing and dies before it has used up all the sugar in the fermentation solution.

Living things make complex molecules far more efficiently than they can be made in a laboratory. For thousands of years people have harnessed microorganisms to make products such as drinks, bread, yoghurt, and cheese. Microorganisms are added to food ingredients and kept in the right conditions. As they grow, by-products from the microorganisms create the desired product. For example, yeast is used in bread-making. It uses sugars in the flour for respiration and the carbon dioxide it produces makes bread rise.

Bioethanol from sugar

Petrol and diesel are used to fuel vehicle engines. They are made from crude oil. **Bioethanol** can also be used to fuel vehicle engines. It is made from the sugars in plant material such as sugar beet, maize, and wheat. Fuel from renewable sources is said to be more **sustainable** and causes less global warming.

Yeast are single-celled organisms. Yeast have been used for thousands of years to brew beers and wine. On a much larger scale, yeast are used to make bioethanol.

Yeast cells take sugars and convert them into ethanol during the process of anaerobic respiration. This process is called **fermentation**. Large fermentation vessels inside the factory contain yeast, sugar, water, and other nutrients. The yeast produce bioethanol that can be used to fuel car engines.

Carbon dioxide, a by-product of the respiration of yeast, has caused these loaves of bread to rise.

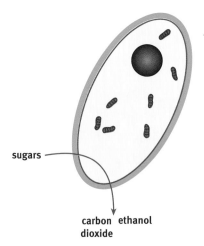

Yeast cells take sugars and convert them into ethanol during the process of fermentation.

sugars

carbon   ethanol
dioxide

This factory produces bioethanol to fuel cars.

Biofuels and sustainability

Biofuels can be controversial. Land may be used to grow crops for fuel instead of food for people. In some areas, forest is cut down so the land can be used to grow crops for fuel. Producers are now developing ways of using non-food crops and algae to end the need to use food crops.

Biogas from waste

Biogas is a fuel obtained from animal manure or human waste. Bacteria break down the organic material in the manure and produce methane gas. This can be used as a fuel to heat buildings and run electricity generators.

Methanogenic bacteria in biogas fermenters produce methane gas during anaerobic respiration.

This chemical factory produces biogas.

Biodigesters typically use manure from farm animals, but some are also used to get biogas from human waste. The biogas contains around 70% methane and 30% carbon dioxide.

Bacteria grow in anaerobic conditions inside the biodigester and produce methane gas, which can be used as a fuel. Manure from animals and human toilets can be collected. The optimum temperature of 36 °C makes this a good option in warm climates.

Biogas is used in some developing countries, such as by smallholders in rural parts of India. It is particularly useful where other sources of energy, such as electricity, are not available.

Key words

- bioethanol
- sustainable
- fermentation
- biogas

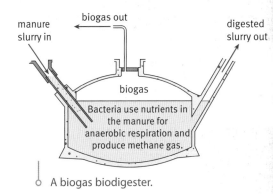

manure slurry in · biogas out · digested slurry out · biogas · Bacteria use nutrients in the manure for anaerobic respiration and produce methane gas.

A biogas biodigester.

Questions

1 Why is bioethanol said to be a renewable resource?

2 Explain why producing and using bioethanol contributes less carbon dioxide to the atmosphere than using petrol from crude oil.

3 Describe the process of anaerobic respiration in yeast. Compare it to anaerobic respiration in animal cells.

4 Biogas contains around 70% methane and 30% carbon dioxide. Suggest where the carbon dioxide in the biogas comes from.

5 Draw a flow diagram to show how energy is transferred from sunlight into the methane in biogas. Include details of the processes happening at each stage.

Science Explanations

All living things are made up of cells. Biological processes such as photosynthesis and respiration take place in cells. These processes involve chemical reactions that are speeded up by enzymes.

You should know:

- that some chemical reactions in cells require energy; these reactions include muscle contraction, synthesis of large molecules, and active transport
- that respiration is a series of chemical reactions in plant, animal, and microbial cells, which release energy by breaking down food molecules
- what enzymes are and how they speed up chemical reactions in living organisms
- that cells make enzymes from the instructions carried in our genes
- why the model that describes how enzymes recognise molecules is called the lock-and-key model
- the conditions that enzymes need to work at their optimum rate
- that photosynthesis uses energy from sunlight to make glucose by joining carbon dioxide and water, and releasing oxygen as a waste product
- that chlorophyll absorbs light energy
- that glucose may be converted into other chemicals needed for plant growth
- how to use equipment such as light meters, quadrats, and identification keys to investigate the effect of light on plants
- how to take a transect in fieldwork
- that minerals are taken up by plant roots, and what they are used for in plants
- that diffusion is the passive overall movement of molecules from a region of higher concentration to a region of lower concentration
- how osmosis is the overall movement of water by diffusion from a dilute to a more concentrated solution through a partially permeable membrane
- why active transport, such as the absorption of nitrates by plant roots, requires energy from respiration
- that aerobic respiration breaks down glucose in the presence of oxygen, releasing energy, carbon dioxide, and water; it takes place in animal and plant cells and some microorganisms
- that anaerobic respiration takes place in animal, plant, and some microbial cells
- how energy is released from glucose without oxygen during anaerobic respiration – animal cells form lactic acid, plant cells and yeast produce carbon dioxide and ethanol
- the structures and functions of typical plant, animal, and microbial cells
- how we use the anaerobic respiration of microorganisms to make biogas, bread, and alcohol.

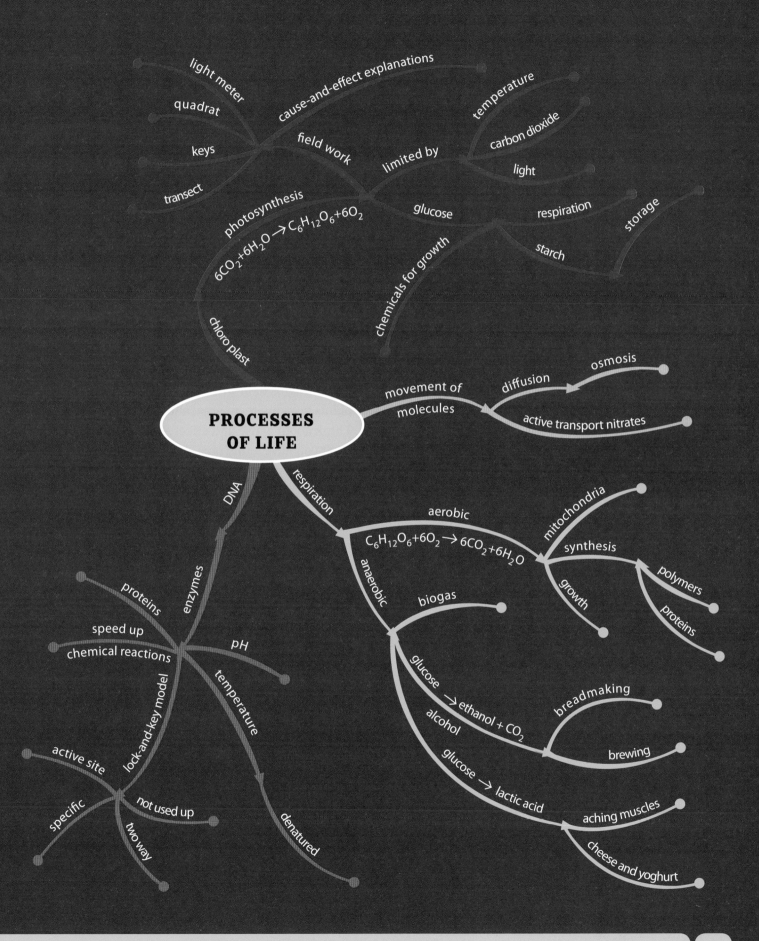

light meter

quadrat

keys

transect

cause-and-effect explanations

field work

photosynthesis

$6CO_2+6H_2O \rightarrow C_6H_{12}O_6+6O_2$

chloroplast

limited by

temperature

carbon dioxide

light

glucose

respiration

storage

starch

chemicals for growth

PROCESSES OF LIFE

movement of molecules

diffusion

osmosis

active transport nitrates

DNA

respiration

aerobic

$C_6H_{12}O_6+6O_2 \rightarrow 6CO_2+6H_2O$

mitochondria

synthesis

growth

polymers

proteins

enzymes

proteins

speed up

chemical reactions

pH

lock-and-key model

temperature

active site

specific

not used up

two way

denatured

anaerobic

biogas

glucose → ethanol + CO$_2$

alcohol

glucose → lactic acid

breadmaking

brewing

aching muscles

cheese and yoghurt

Ideas about Science

This module helps you develop an understanding of life processes. You will also learn more about how scientists explain cause and effect, and how they investigate relationships between factors.

If you take several measurements of the same quantity, these are likely to vary. This may be because:
- you are measuring several individual examples, such as oxygen produced from different samples of pond weed
- the quantity you are measuring is varying, for example, the number of bluebells growing in different areas of a wood
- the limitations of the measuring equipment or because of the way you use the equipment.

The best estimate of a true value of a quantity is the mean of several measurements. The true value lies in the spread of values in a set of repeat measurements.
- A measurement may be an outlier if it lies outside the range of the other values in a set of repeat measurements.
- When comparing information on plants growing in different places, a difference between their means is likely to be real if their ranges do not overlap.
- A correlation shows a link between a factor and an outcome, for example, as light intensity increases, the rate of photosynthesis increases.
- A correlation does not always mean that the factor being changed causes the outcome.

Scientists often think about processes in terms of factors that may affect an outcome (or outcome variable). When you investigate the relationship between a factor and an outcome, it is important to control all the other factors that you think might affect the outcome of the investigation. This is called a 'fair test'.

You should be able to:
- identify input variables for photosynthesis
- explain how temperature and pH affect enzyme reactions.

You also need to explain why it is necessary to control all of these variables in an investigation and consider how you would perform a fair test.

When you plan an investigation you need to identify the likely effect of a factor on an outcome. You also need to know that if you don't control the variables in an investigation, the investigation will be flawed.

You should understand why plants grow in particular places. You could measure factors like soil pH, temperature, light intensity, and water. Compare this data with what you know about the plants growing in an area, and what plants need. Think about cause-and-effect explanations for what you have found out.
- Enough individual samples must be taken to get a true picture of the conditions.

You should be able to find ways of collecting reproducible data and be able to explain why repeating measurements leads to a better estimate of the quantity measured. Here are two examples:
- A light meter can be used to accurately measure light intensity. The sensor should be held at the same angle for each reading. Time of day and shading from trees or bushes should be taken into account.
- One way of sampling using quadrats is to position quadrats randomly in the area being measured. This removes bias. Recording the quadrat data involves accurate identification of each species.

Review Questions

1 This question is about enzymes.

a What are enzymes?
Your answer should include:
- what type of chemical they are
- what they do.

b Enzymes and some molecules fit together.
Which model explains this?
Choose the correct answer.

> **enzyme-and-molecule model**
> **lock-and-key model**
> **puzzle-shaped model**

c Which of the following does not affect biological enzymes?
- **i** light
- **ii** pH
- **iii** temperature

2 Alex draws a set of diagrams to show what happens during a reaction involving an enzyme.
He makes a mistake and draws one incorrect stage.

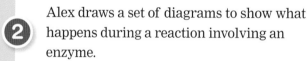

The stages are not drawn in the correct order.
Put the stages in the correct order.
The last one has been done for you.

			D

3 Photosynthesis takes place in green plants.

a Describe three factors that limit the rate of photosynthesis.

b The rate at which photosynthesis takes place can be measured. One way to do this is to count the number of bubbles given off by a water plant in one minute.

light source

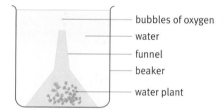
bubbles of oxygen
water
funnel
beaker
water plant

Explain why trying to count the number of bubbles in one minute may not give an accurate value for the rate of photosynthesis.

4 Steve is an athlete. He runs marathons. As he runs he releases energy both aerobically and anaerobically.
Explain **how** and **why** Steve uses both of these types of respiration when he runs a marathon.

5 Molecules move across cell membranes by diffusion, osmosis, and active transport.
Describe and explain the differences between these three different processes.

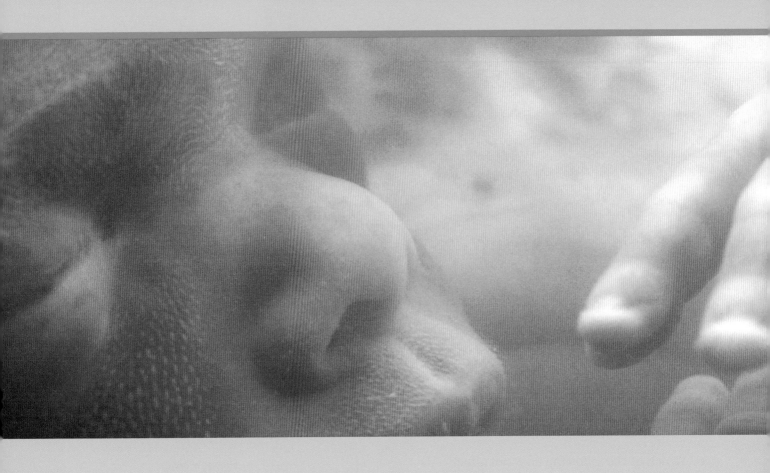

B5 Growth and development

Why study growth and development?

How does a human embryo develop? What makes cells with the same genes develop differently? Exploring questions like these is part of the fast-moving world of modern biology.

What you already know

- Genes affect the way organisms including humans develop.

- Clones such as plants grown from cuttings are more similar to each other than organisms with a combination of their parents' genes.

- Organisms are made up of cells, and these divide and develop into the whole organism.

Find out about

- the structure of DNA, and how it controls the proteins a cell makes

- how cells divide to make sex cells and to make new body cells

- how cells become specialised

- the differences between plant and animal growth.

The Science

Sex cells carry the genetic information to make a new individual. Cells become specialised because of the different proteins they make. DNA is the chemical that genes are made of, and its unique structure determines the proteins a cell makes.

Ideas about Science

Data and explanations of data are different. Creative thinking is needed to develop explanations from data.

Find out about

- different cells, tissues, and organs
- growing up

You began life as a single cell. By the time you were born you were made of millions of cells. You probably weighed about 3 or 4 kilograms. Now you probably weigh over 50 kilograms. Not only have you grown, but you have changed in many ways. In other words, you have developed.

Since you were born, your **development** has been gradual. In some plants and animals, development involves big changes. For example, the young and the adults in these photographs look very different.

Questions

1. Match the young and the adults in pictures A to F.

2. Which animal in the pictures has a life cycle most like a fly?

Building blocks

Like you, the plants and animals in the pictures above are multicellular. Not every cell is the same. Your body has more than 300 different kinds of cell. Each kind of cell is **specialised** to do a particular job. Muscle cells can contract to cause movement, nerve cells are specially shaped to carry impulses, and red blood cells lack nuclei, giving extra room to carry oxygen.

Tissues and organs

All newly formed human cells look much the same. Then they develop into groups of specialised cells called **tissues**. Our body has muscle tissue, nervous tissue, and fatty tissue. Bone, cartilage, and blood are also tissues.

Plant cells are different from animal cells, but they are specialised too. Plant cells have cell walls outside the cell membrane, and some have spaces called vacuoles.

As an animal embryo or plant grows, groups of tissues arrange themselves into **organs**, for example, the heart and brain in humans, and roots, leaves, and flowers in plants. The heart is made up of muscle tissue, nervous tissue, connective tissue, and some fatty tissue. The diagram below shows the different plant tissues found in a leaf, which is a plant organ.

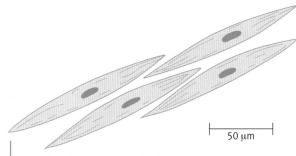

50 µm

Muscle cells contract and relax to cause movement.

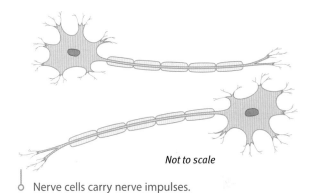

Not to scale

Nerve cells carry nerve impulses.

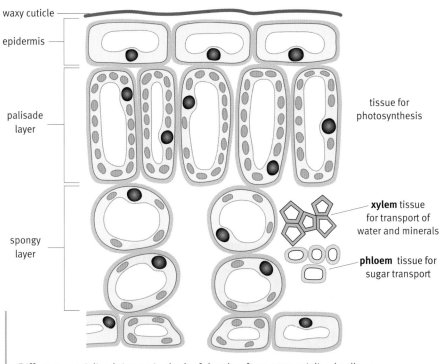

waxy cuticle

epidermis

palisade layer

tissue for photosynthesis

spongy layer

xylem tissue for transport of water and minerals

phloem tissue for sugar transport

Different specialised tissues in the leaf develop from unspecialised cells.

Key words
- ✔ development
- ✔ specialised
- ✔ tissues
- ✔ organs
- ✔ xylem
- ✔ phloem

Question

3 Explain the difference between a tissue and an organ.

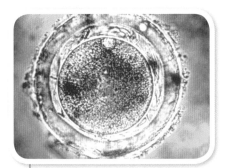

A fertilised human egg cell.

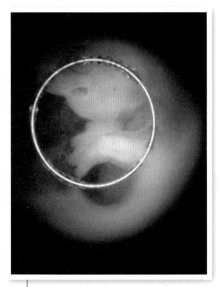

A human fetus at eight weeks. This photograph was taken from inside the mother's uterus. The fetus is about 2.5 cm long.

Key words

- ✓ **zygote**
- ✓ **clone**
- ✓ **embryonic stem cells**
- ✓ **meristem cells**
- ✓ **fetus**

Questions

4   What is a zygote?

5   When does a human embryo become a fetus?

From single cell to adult

All the cells in your body come from just one original cell – a fertilised egg cell or **zygote**.

So the zygote must contain instructions for making all the different types of cells in your body, for example, muscle cells, bone cells, and blood cells. It also has the information to make sure that each type of cell develops in the right place and at the right time. This information is in your DNA – the chemical that your genes are made of.

In humans:

sperm + egg cell →(fertilisation)→ zygote (fertilised egg cell) →(division of cells by mitosis)→ embryo

The growing baby

During the first week of growth, the zygote divides by mitosis to form a ball of about 100 cells. The nucleus of each cell contains an exact copy of the original DNA. As the embryo grows, some of the new cells become specialised and form tissues. After about two months, the main organs have formed and the developing baby is called a **fetus**. A six-day embryo is made of about 50 cells. Adults contain about 10^{14} cells, each with the DNA exactly copied.

When the embryo is a ball of eight cells or fewer, it occasionally splits into two. A separate embryo develops from each section. When this happens, identical twins are produced. They are **clones** of each other. This shows that there are cells in the early embryo that are identical and unspecialised. These cells can develop into complete individuals, and are called **embryonic stem cells**. You can read more about these cells in Section J.

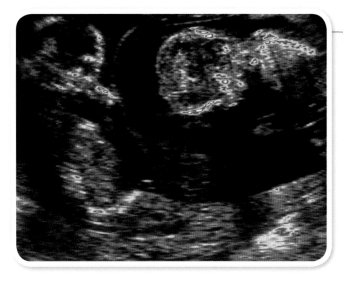

This ultrasound scan shows that twins are expected. Both of the babies' heads can be seen.

Growth patterns

For living things to grow bigger, some of their cells must divide to make new cells. You will probably stop growing taller by the time you are about 18–20 years old.

Flowering plants continue to grow throughout their lives.
- Their stems grow taller.
- Their roots grow longer.
- To hold themselves upright, most stems increase in girth or have some other means of support.

Plants increase in length by making new cells at the tips of both shoots and roots. They also have rings of dividing cells in their stems and roots to increase their girth. These dividing cells are called **meristem cells**.

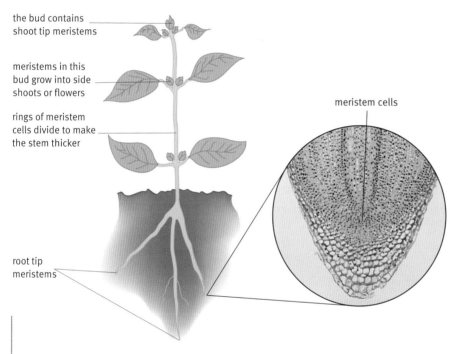

the bud contains shoot tip meristems

meristems in this bud grow into side shoots or flowers

rings of meristem cells divide to make the stem thicker

root tip meristems

meristem cells

Meristem cells divide to make stems and roots longer and make the stem thicker. On the right is a root tip meristem (photographed through a microscope).

This giant sequoia tree is over 2000 years old, 83 m tall, and 26 m in girth (circumference). It is known as 'General Sherman' and is officially the largest giant sequoia tree and the largest living thing on Earth.

Questions

6 Why is it important that all living things have cells that can divide?

7 Name the type of cell in a plant that can divide to make new tissues.

8 Explain how plants:
   a grow taller
   b grow longer roots
   c grow thicker in girth.

Honeysuckle twines its stem around other plants for support.

Find out about

✓ **why plants are so good at repairing damage**

① normal front leg

② leg bitten off by a predator

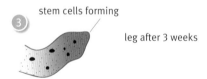
③ stem cells forming

leg after 3 weeks

④ leg after 4 weeks

⑤ leg after 6 weeks

⑥ leg after 10 weeks

If a newt's limb is bitten off by a predator, it can grow a replacement. Most animals can only make small repairs to their body.

Cells in your body divide when you are growing. If you cut yourself, cells divide to repair your body, but only for small repairs. Many plants and some animals can replace whole organs.

Why can plants grow back?

Plant meristem cells are **unspecialised**. Plants keep some meristem cells all through their lives. These are spare back-up cells that divide to make any kind of cell the body needs. So plants can make new xylem or phloem tissues, or can regrow whole organs, such as leaves, if they are damaged. Scientists who develop genetically modified crops can put this to use. They can regenerate whole plants from small pieces of unspecialised plant tissue that they have modified.

How do newts grow?

Animals also have spare back-up cells called **stem cells**. These cells divide, grow, and develop into specialised cells that the body needs.

Animal growth

Newts' stem cells stay unspecialised throughout their lives. So newts can grow new legs if they need to – or even an eye.

The stem cells in adult humans are not as useful, because they have already started to specialise. For example, the stem cells in your skin can only develop into skin cells.

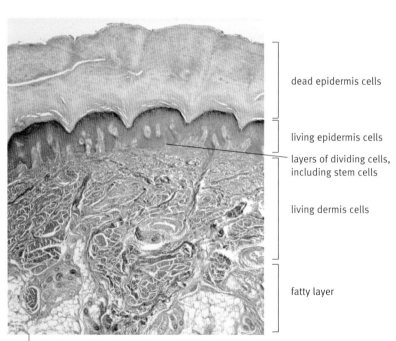

dead epidermis cells

living epidermis cells

layers of dividing cells, including stem cells

living dermis cells

fatty layer

A cross section through human skin. Some of the stem cells continue to grow and divide. Others replace skin cells at a wound or those that wear off at the surface. (Magnification ×36.)

You replace millions of skin cells every day. Most of the dust in your home is worn-off skin cells. Other tissues in your body need a constant supply of new cells. For example, bone marrow contains stem cells to make new blood cells.

Using meristems to make more plants

In module B1, you saw that gardeners use meristems when they grow new plants by taking **cuttings**. Cuttings are just shoots or leaves cut from a plant. In the right conditions they develop roots and grow into new plants.

Some cuttings grow new roots when you put them in water or compost. Others grow better when you dip the cut ends in **rooting powder** before you plant them. Rooting powder contains plant hormones called **auxins**. Auxins cause the new cells produced by the base of the shoot meristem to develop into roots.

By taking cuttings, gardeners can produce lots of new plants quickly and cheaply. But this is not the only reason that they do it. All the cuttings taken from one plant have identical DNA. As they are genetically identical, they are called clones. So taking cuttings is a good way of reproducing a plant with exactly the features that you want. When flowering plants produce seeds, they are reproducing sexually. So new plants grown from seeds vary. They are not identical.

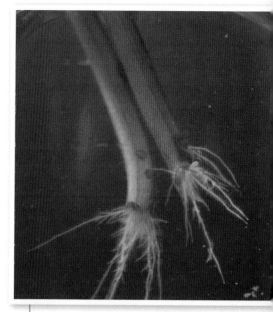

This willow cutting needed only water to grow.

Rooting powder.

Questions

1 For each of these types of cell, say whether they are fully unspecialised or not:
   a meristem cells
   b adult human stem cells.

2 Explain why a newt can regrow a leg but a human cannot.

3 Give two reasons for growing plants from cuttings.

4 Explain how rooting powder helps a plant cutting to grow.

5 A gardener wants to grow dahlias with a variety of colours and sizes. Should she grow them from cuttings or seeds? Explain your answer.

Key words
- ✓ **unspecialised**
- ✓ **stem cells**
- ✓ **cuttings**
- ✓ **auxins**
- ✓ **rooting powder**

Find out about

✓ **why plants grow towards light**

Plants rooted in soil cannot move from place to place – not even the 'walking palm' tree in the picture below. Plant seeds can land anywhere. To survive, the growing plant will need to get enough light for photosynthesis. The plant needs to sense and respond to the environment. In nature, plants continually compete with each other in an environment that is constantly changing.

This houseplant has grown towards the window to increase the amount of light falling on its leaves.

The walking palm tree, *Socratea durissima*, in Costa Rica. New roots grow towards a sunny patch and pull the stem and leaves towards the light. Hormones control the direction of root growth. Older roots in the shade die.

You may have noticed that plants on windowsills seem to bend towards the light. They are not moving, but growing. The direction the light comes from affects the direction of plant growth. This is called **phototropism**. Phototropism increases a plant's chances of survival.

Questions

1 Write a definition for phototropism.

2 How does the walking palm tree grow towards the light?

3 Explain why a plant benefits from bending towards the light.

Darwin's phototropism experiments

Charles Darwin experimented with phototropism. He showed that the young shoots of grasses:

- normally grew towards light
- remained straight when he covered their tips.

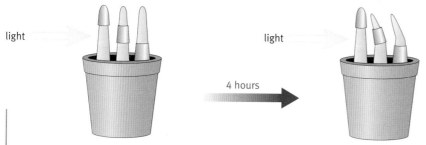

light

light

4 hours

○ Foil covers different parts of the barley shoots.

In the experiment, shown in the picture above, covering the lower parts of the shoot did not stop bending towards the light. This shows that only the tip is sensitive to light. The shoot bends below the tip – where cells are no longer dividing but are increasing in length.

Darwin did not know how bending towards light happened but his results allowed him to explain which parts of the plant sense and respond to light. Now scientists have found that higher concentrations of plant hormones called auxins cause shoot cells to expand. The diagrams on the right explain how this causes phototropism.

Questions

4 Look at the diagram showing phototropism experiments in barley. Suggest where:
   a the shoot detects light
   b the cells are growing very quickly to cause bending.

5 Explain why Darwin could not develop a full explanation from the data he collected on phototropism.

6 Distinguish between hypothesis, theory, explanation, and data in the experiment on phototropism.

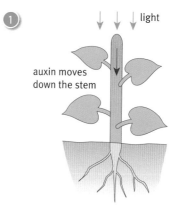

① light

auxin moves
down the stem

When a growing shoot gets light from above, the auxins spread out evenly and the shoot grows straight up.

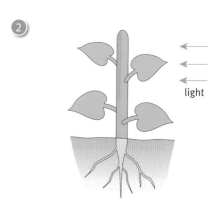

②

light

When light comes from one side, the auxins move over to the shaded side.

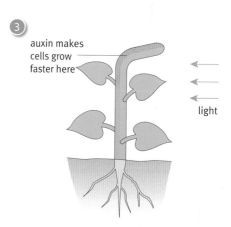

③ auxin makes
cells grow
faster here

light

The shoot grows faster on the shaded side, making the shoot bend towards the light.

○ How auxins explain phototropism.

Key word

✓ **phototropism**

Find out about

✔ **where genes are kept inside your cells**

All cells start their lives with a nucleus. A few specialised cells lose their nuclei when they finish growing. Human red blood cells are one example. Their job is to carry oxygen attached to haemoglobin molecules.

Red blood cells develop from stem cells in your bone marrow. As they develop, they make more and more haemoglobin. By the time they leave the bone marrow, they are full of haemoglobin and their nuclei have broken down.

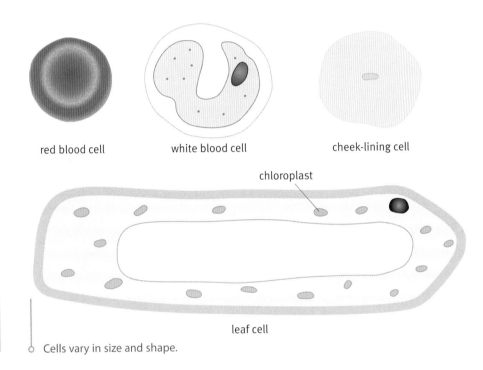

red blood cell   white blood cell   cheek-lining cell

chloroplast

leaf cell

Cells vary in size and shape.

Organism	Estimated gene number	Chromosome number
human	~30 000	46
mouse	~30 000	40
fruit fly	13 600	8
Arabidopsis thaliana (plant)	25 500	5
roundworm	19 100	6
yeast	6 300	16
Escherichia coli (bacterium)	3 200	1

Chromosomes

A **chromosome** is a long molecule of DNA wound around a protein framework. You have about a metre of DNA in each of your nuclei. This is made up of about 30 000 **genes**. Each gene codes for a protein or part of a protein.

Different species have different numbers of chromosomes and different numbers of genes (see the table on the left).

You have 23 pairs of chromosomes in your nuclei. You got one set of 23 from your mother's egg cell nucleus and the other set from the nucleus of your father's sperm.

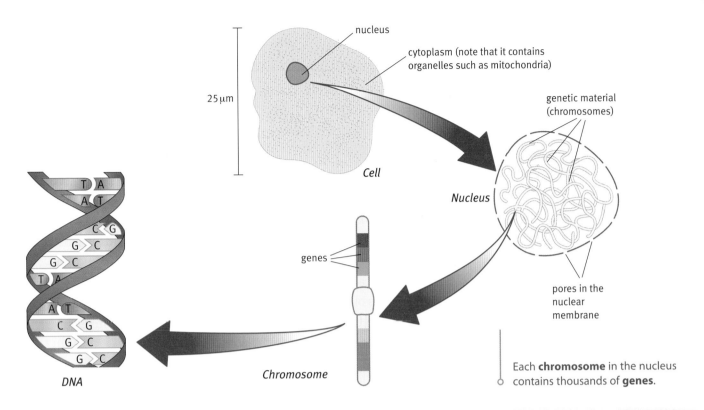

nucleus

cytoplasm (note that it contains organelles such as mitochondria)

25 μm

Cell

genetic material (chromosomes)

Nucleus

genes

pores in the nuclear membrane

Chromosome

DNA

Each **chromosome** in the nucleus contains thousands of **genes**.

These people are more alike than it appears. 99.9 % of their genes are the same.

What is special about DNA?

You will find out later in the module that the molecule of DNA has a particular structure that allows it to:

- make exact copies of itself
- provide instructions so that the cell can make the right proteins at the right time.

Key words
- ✓ **chromosomes**
- ✓ **genes**

Questions

1 Name two ways in which your red blood cells are different from the other cells in your body.

2 Suggest why red blood cells wear out and have to be replaced every 2–3 months.

3 How many different types of protein are there likely to be in yeast?

4 Which organism has half as many genes as yeast?

5 What two properties does DNA have that make it work as genetic material?

Find out about

✓ **how your cells divide for growth and to repair your body**

Imagine a space probe bringing back objects like this from Mars. Scientists would need to find out whether they were alive or not. It might be a living organism able to colonise Earth!

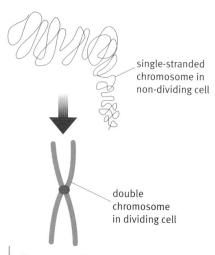

single-stranded chromosome in non-dividing cell

double chromosome in dividing cell

You can see chromosomes only in dividing cells.

Key words

✓ **organelles**
✓ **mitosis**
✓ **ribosomes**
✓ **mitochondria**

It is not always easy to tell whether something is a living organism or not.

You could ask yourself:

• can it grow and reproduce?
• is it made of cells?

Life cannot exist without the growth, repair, and reproduction of cells.

Cell division

When new body cells are made, they contain the same number of chromosomes as each other and the parent cell. They also contain the same cell parts, called **organelles**. So, before a cell divides, it must grow and make copies of:

• other **organelles** – such as **ribosomes** and **mitochondria**
• its nucleus, including the chromosomes.

Only then does the cell divide. This part of the process is called **mitosis**.

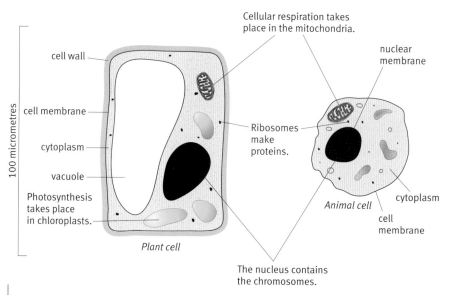

Cellular respiration takes place in the mitochondria.

nuclear membrane

cell wall

cell membrane

cytoplasm

vacuole

Photosynthesis takes place in chloroplasts.

100 micrometres

Ribosomes make proteins.

Animal cell

cytoplasm

cell membrane

Plant cell

The nucleus contains the chromosomes.

Cell organelles, such as mitochondria, are copied before a cell divides.

Copying chromosomes

You can see the chromosomes in dividing cells by using a light microscope. The DNA is too spread out in other cells to be visible. After the chromosomes are copied, the DNA strands become shorter and fatter. You can read more about how DNA is copied in Section G.

Mitosis

During mitosis, copies of chromosomes separate and the whole cell divides.

First, a complete set of chromosomes goes to each end of the dividing cell and two new nuclei are formed. A complete set of organelles also goes to each end. Then the cytoplasm divides to form two identical cells. In plant cells, a new cell wall forms too.

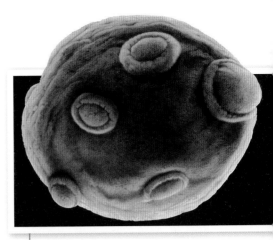

In yeast, new cells grow as buds from the parent.

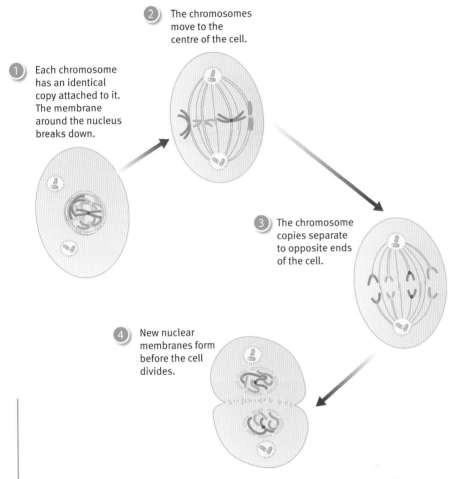

2 The chromosomes move to the centre of the cell.

1 Each chromosome has an identical copy attached to it. The membrane around the nucleus breaks down.

3 The chromosome copies separate to opposite ends of the cell.

4 New nuclear membranes form before the cell divides.

The pictures show what happens in mitosis in an animal cell. They show only four chromosomes.

Daffodil bulbs divide to form new ones.

Mitosis and asexual reproduction

Some plants and animals reproduce asexually. They use mitosis to produce cells for a new individual.

This means that each of the individuals produced in asexual reproduction is genetically identical to the parent, so it is a clone of its parent.

Questions

1 a How many cells are made by mitosis?
  b What are these new cells like compared with their parent cell?

2 What might scientists do to find out whether the objects from Mars in the cartoon opposite are living things or not?

3 Before a cell divides, it grows. What two steps happen during cell growth?

4 What two main steps happen during cell division by mitosis?

Find out about

✓ **cell division to make gametes**

Most plants and animals reproduce sexually. Males and females make sex cells or **gametes**, which join up at fertilisation. The fertilised egg, or zygote, develops into the new life.

It is not always obvious which plant or animal is male and which is female. Some plants and animals are both.

Only the male peacock has magnificent tail feathers.

The berries show that the holly on the right is female. You cannot be sure about the one on the left.

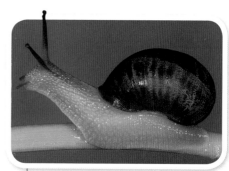

A snail has both male and female sex organs.

The only way to be sure about the sex of an organism is to look at its gametes. Males have small gametes that move. Females have large gametes that stay in one place.

Human males produce sperm in their testes. Females produce egg cells in their ovaries.

250 micrometres

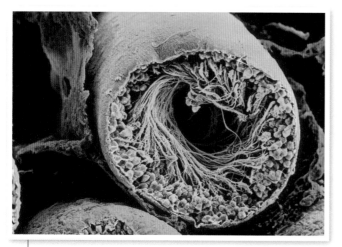

Sperm develop in tubules in the testes.

600 micrometres

Pollen contains the male gametes of a flowering plant.

Male gametes are usually made in very large numbers. They move to the female gamete by swimming or being carried by the wind or an insect.

What is special about gametes?

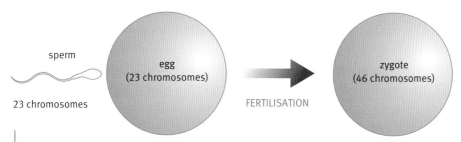

Meiosis halves the number of chromosomes in gametes. Fertilisation restores the number in the zygote.

Human body cells have 23 pairs of chromosomes – 46 in total. Gametes have only 23 single chromosomes. This is important because when a sperm cell fertilises an egg cell, their nuclei join up. The fertilised egg cell (zygote) gets the correct number of chromosomes: 23 pairs – 46 in total. Half your chromosomes come from your mother and half come from your father. Gametes are made by a special kind of division called **meiosis**.

In humans, meiosis makes gametes that:
- have 23 single chromosomes (one from each pair)
- are all different – no two gametes have exactly the same genetic information.

Offspring from sexual reproduction are different from each other and from their parents. We say that they show **genetic variation**.

Meiosis

Meiosis starts with normal body cells. It only happens in sex organs.

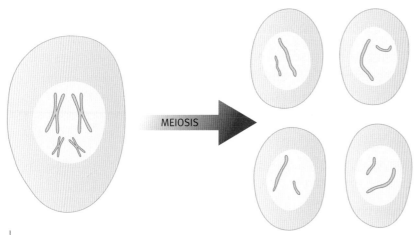

The chromosomes in this diagram have been copied. The parent cell divides twice, producing four cells. There are four cells after meiosis. They have half the number of chromosomes as the parent cell.

Questions

1 Look again at the photos of holly at the top of the opposite page. You can be sure that the holly on the right is female. Why can you not be sure about the sex of the holly on the left?

2 Why is it important that gametes have only one set of chromosomes?

3 Why are male gametes made in such large numbers?

Find out about

- the structure of DNA
- how DNA is copied for cell division

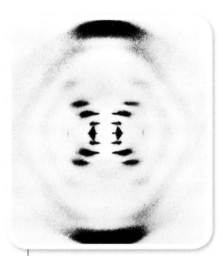

X-ray diffraction pictures of DNA, like this, show a repeating pattern.

Watson (left) and Crick reveal their model of DNA.

In 1865 Gregor Mendel published his work on pea plants. You learnt about this in module B3. Mendel's data showed how features could be passed on from parents to their young. The explanation for how information is passed on did not simply emerge from Mendel's data. It took the work and creative thinking of many scientists to explain how it happens.

1859	A chemical was extracted from nuclei and named 'nuclein'.
1944	'Nuclein' was recognised as genetic material.
Late 1940s	Erwin Chargaff discovered a pattern in the number of bases in DNA.
1951	Linus Pauling and Robert Corey showed that proteins have a helix structure.
1952–3	Rosalind Franklin and Maurice Wilkins produced X-ray diffraction pictures of DNA. They showed that the molecule had a regular, repeating structure.

Some of the discoveries that led up to the discovery of DNA structure. These discoveries provided explanations that accounted for Mendel's data. Predictions based on the explanations were then tested by further experiments.

Solving the mystery

In 1953, Francis Crick and James Watson published their now famous paper 'A Structure for Deoxyribose Nucleic Acid', in the scientific journal *Nature*. Their paper brought together all the work done on DNA. They used it to work out the **double-helix** structure of DNA.

Base pairing

There are four special molecules called 'bases' in DNA: adenine (A), thymine (T), guanine (G), and cytosine (C).

Erwin Chargaff had discovered that the amount of A is always the same as the amount of T, and the amount of G is the same as the amount of C. This is true no matter what organism the DNA comes from.

Crick and Watson concluded from this evidence that:
- A always pairs with T
- G always pairs with C.

This is **base pairing**.

The double helix

Watson made cardboard models of the bases. He found that A+T was the same size as G+C. Suddenly, he realised what that meant. In a molecular model, he and Crick fitted the pairs of bases between two chains of the other chemicals in DNA, a sugar called deoxyribose and phosphates. It worked. The shape turned out to be a double helix – a bit like a twisted ladder. Better still, it matched the X-ray evidence.

How DNA passes on information

Watson and Crick's structure not only accounted for what was known about DNA so far, it also let them predict how DNA might copy itself exactly. Base pairing means that it is possible to make exact copies of DNA:

- Weak bonds between the bases split, unzipping the DNA from one end to form two strands.
- Immediately, new strands start to form from free bases in the cell.
- As A always pairs with T, and G always pairs with C, the two new chains are identical to the original.

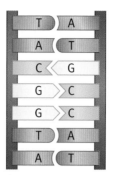

The rungs of the ladder are the pairs of bases held together by weak chemical bonds.

There are ten pairs of bases for each twist in the helix.

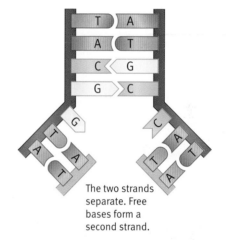

The two strands separate. Free bases form a second strand.

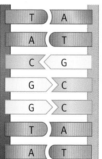

Each DNA molecule is made of half old DNA (black) and half new DNA (grey).

Each half of the split DNA molecule is complete, making two identical DNA molecules.

The structure of DNA (schematic).

Questions

1 The shape of a DNA molecule is sometimes described as a twisted ladder.
   a What makes the sides of the ladder?
   b What part of the ladder are the bases?

2 The bases in a DNA molecule always pair up the same way. Which base pairs with:
   a A?   c G?
   b C?   d T?

3 Describe what happens when a DNA molecule is copied.

4 Which observations made by other scientists did Watson and Crick's model account for?

5 Which aspect of Watson and Crick's work could be described as 'creative thinking' rather than making simple deductions from the data?

Key words
- ✓ double helix
- ✓ base pairing

Find out about

✓ **how DNA controls which protein a cell makes**

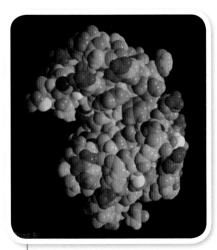

Enzymes work because of the shape of their active sites.

The great number of jobs carried out by proteins means that they are very different from one another. The exact shape of a protein can be very important to how it works. Cells make proteins from about 20 different **amino acids**. Chains of between 50 and many thousands of amino acids are formed. In each protein, the amino acids are joined in a particular order, but there are thousands of possibilities. In the same way we can make thousands of different words using only 26 letters in our alphabet. The order of the amino acids fixes the way the chains of amino acids fold to form the three-dimensional shape of the protein.

The genetic code

In 1961, Crick worked out that three bases on a DNA strand code for each amino acid. This is called a **triplet code**. Different combinations of the four bases (A, T, G, and C) produce 64 triplet codes. So there is more than one code for each amino acid. Crick guessed this because any fewer bases would not give enough codes for every amino acid. There are also codes for start and stop. They mark the beginning and end of a gene. Subsequent experiments have demonstrated that Watson and Crick's explanations were correct. Their work forms the foundation of the science of genetic engineering.

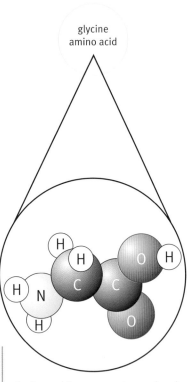

Amino acids are complex molecules. The diagrams on this page and the next represent the molecule simply.

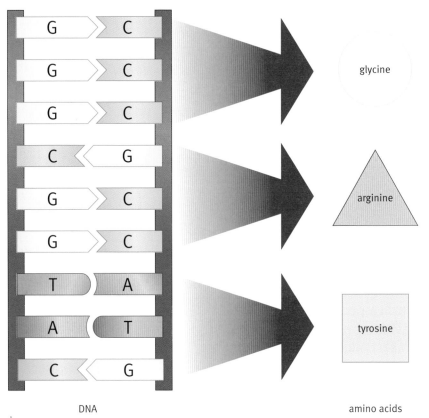

Three bases on the DNA code for each amino acid. For example, TTT codes for lysine.

Which part of a cell makes proteins?

DNA in the nucleus contains the genetic code for making the proteins. The proteins are made on tiny organelles in the cytoplasm, called ribosomes. Genes cannot leave the nucleus. So how do ribosomes get the instruction for making a protein? A molecule small enough to get through the pores of the nuclear membrane transfers the genetic code to the ribosomes. This smaller molecule is called messenger RNA (mRNA).

The differences between DNA and mRNA are that mRNA has:

*   only one strand
*   the base uracil (U) in place of the thymine (T) in DNA.

The diagram below shows how a protein is made.

① The gene unzips, and mRNA bases pair with DNA bases to form a strand of mRNA. U matches up with A instead of T.

DNA   DNA   mRNA

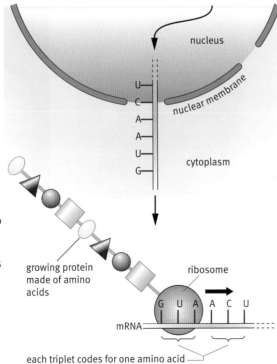

② The RNA moves out of the nucleus to one of the many ribosomes in the cytoplasm.

nucleus

nuclear membrane

cytoplasm

③ The ribosome attaches to one end of the mRNA. As it moves along the mRNA, the ribosome reads the genetic code so that it can join the amino acids together in the correct order. When it has finished, the ribosome releases the protein into the cytoplasm and starts to make another one.

growing protein made of amino acids

ribosome

mRNA

each triplet codes for one amino acid

DNA makes protein with a ribosome's help (schematic).

Questions

1   Why are instructions for making proteins copied onto mRNA?

2   How many DNA bases code for each amino acid?

3   What is the DNA triplet code for tyrosine?

4   Which amino acid has the DNA code GCC?

5   Make bullet-point notes to explain how a protein is made. Your first bullet point should be:
    *   the gene unzips.

    Your last bullet point should be:
    *   the ribosome releases the protein into the cytoplasm.

Find out about

✔ **some different proteins in your body**
✔ **why cells become specialised**

An oak tree has about 30 different types of cell. Your body has more than 300 types of cell. Each cell type has its own set of proteins.

Some proteins make up the framework of cells and tissues. These are **structural proteins**. If we take away all the water in an animal cell, 90% of the rest is proteins.

Protein	Found in …	Property
keratin	hair, nails, skin	strong and insoluble
elastin	skin	springy
collagen	skin, bone, tendons, ligaments	tough and not very stretchy

Different structural proteins have different properties.

The flesh of meat and fish is the animals' muscles. Muscles are mainly protein and provide the protein in many people's diets. Plant seeds such as soya and other beans are rich in proteins and are the basis of many vegetarian meals.

Other proteins are essential for the chemical reactions that keep our bodies working. For example, **enzymes** speed up the chemical reactions in a cell. **Antibodies** are the proteins that help to defend us against disease.

Questions

1 Name three types of protein in your body.

2 Name one structural protein and say how it is suited to do its job.

An oak tree has about 30 types of cell.

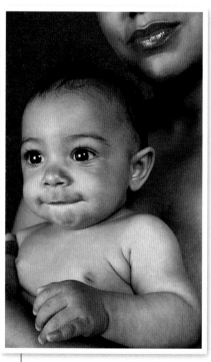

Humans have over 300 different types of cell.

What is the link between genes and proteins?

All the cells in your body come from just one original cell, the zygote. This divides to form a ball of cells. Soon, cells start to specialise. They make the proteins needed to become a particular type of cell.

DNA is a cell's genetic code. Each gene is the instruction for a cell to make a different protein. By controlling what proteins a cell makes, genes control how a cell develops.

Each of your cells has a copy of all your genes. Something must make some cells turn into nerve cells and others into heart cells and all the other cell types. There must be **genetic switches**, but how they work is still being researched by scientists.

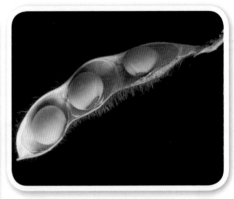

Rhino horn, tortoise shell, soya beans, steak … lots of different proteins.

Questions

3 DNA contains a cell's genetic code. What does DNA do in a cell?

4 How do genes control how a cell develops?

5 Where would you find cells that make these proteins:
   a collagen?
   b amylase?
   c haemoglobin?

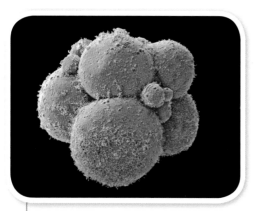

This is one chromosome from a salivary gland of a midge. The green areas are the active genes where DNA unravels whilst the protein is being made. One of these is instructing the cell to make a lot of amylase.

Up to the eight-cell stage, each cell of a human embryo can develop into any kind of cell – or even into a whole organism.

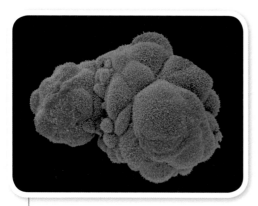

At six days and about 50 cells, the cells in this human embryo are already specialised and will form different types of tissue. The different cells can no longer develop into a whole organism.

Gene switching

Each gene controls the manufacture of one type of protein. So, in any organism, there are as many genes as there are different types of protein. In humans there are 20 000 to 25 000 genes.

Not all these genes are active in every cell. As cells grow and specialise, some genes switch off.

In a hair cell, the genes for the enzymes that make keratin will be switched on:

hair cell genes → enzymes for → hair
switched on   making keratin   grows

But the genes for those that make amylase will be switched off.

In a salivary gland cell:

salivary gland cell → amylase → starch
genes switched on   secreted   digested

Gene switching in embryos

An early embryo is made entirely of embryonic stem cells. These cells are unspecialised up to the eight-cell stage. All the genes in these cells are switched on. As the embryo develops, cells specialise. Different genes switch off in different cells.

unspecialised cell

All the genes are switched on in this chromosome.

hair cell

salivary gland cell

Different genes are switched off in specialised cells.

Key

gene switched on   gene switched off

Some proteins are found in each type of cell, for example, the enzymes needed for respiration. All cells respire, so the genes needed for respiration are switched on in all cells.

In adults, there are stem cells in parts of the body where there is regular replacement of worn out cells. These can only develop into cells of the particular organ where they are positioned. So some of their genes must be switched off.

Right cell, right place

Compare the fingers on your right hand with the same fingers on your left hand. They are probably almost mirror images of each other. As we grow, the position and type of cells must be controlled, so each tissue and organ develops in the right place.

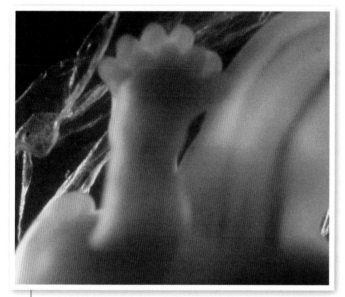

Cells near the end of a limb will make fingers. Cells nearer the body will make the arm. This happens because of the difference in the concentrations of chemical signals in each region of the embryo.

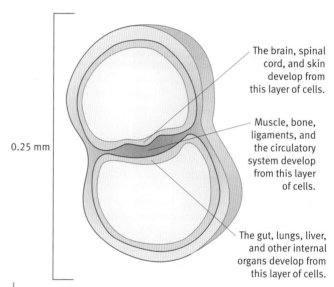

The brain, spinal cord, and skin develop from this layer of cells.

Muscle, bone, ligaments, and the circulatory system develop from this layer of cells.

The gut, lungs, liver, and other internal organs develop from this layer of cells.

0.25 mm

It is possible to map specialised parts of the body onto a diagram of a 14-day-old human embryo. Here you can see which groups of cells in the embryo will develop into future tissues and organs.

Questions

6   Suggest a function, other than respiration, that all cells carry out.

7   At the eight-cell stage of any embryo, how many genes are switched on?

8   What is the evidence that some genes are switched off at the 50–100-cell stage?

Find out about

✔ **scientific research to use stem cells for treating some diseases**

Scientists grew this skin from skin stem cells in sterile conditions. Doctors use it for skin grafts.

Key word

✔ **therapeutic cloning**

A lot of research into stem cells is currently going on. This is because many scientists see the possibility of using them for:

- the treatment of some diseases
- the replacement of damaged tissue.

Imagine if scientists could produce …	They might use them to treat …
nerve cells	Parkinson's disease and spinal cord injuries
heart muscle cells	damage caused by a heart attack
insulin-secreting cells	diabetes
skin cells	burns and ulcers
retina cells	some kinds of blindness

The problem is to find stem cells of the correct type and then to grow them to produce enough cells. Stem cells can come from early embryos, umbilical cord blood, and adults. Embryonic stem cells are the most useful because the cells are not yet specialised. All their genes are still switched on up to the eight-cell stage.

There are problems with this new technology, which is called **therapeutic cloning**. For example, tissues from the embryonic stem cells do not have the same genes as the person getting the transpant. Transplanted tissue is rejected if your body recognises that the cells are not from your body.

Cloning from your own cells

Another possibility is to remove the nucleus of a zygote and replace it with the nucleus from a patient's own body cell. The new embryo would have the same genes as the patient. So the embryonic stem cells produced match those of the patient.

Therapeutic cloning.

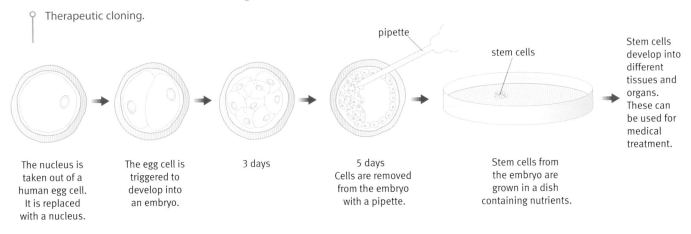

The nucleus is taken out of a human egg cell. It is replaced with a nucleus.

The egg cell is triggered to develop into an embryo.

3 days

5 days
Cells are removed from the embryo with a pipette.

Stem cells from the embryo are grown in a dish containing nutrients.

Stem cells develop into different tissues and organs. These can be used for medical treatment.

Using adult stem cells

A way to avoid using embryos would be use to use the 'adult' stem cells already in your body. Adult bone marrow stem cells from a donor are routinely used to treat leukaemia (cancer of the blood). The stem cells in the transplant restore healthy blood-cell production in the recipient's own bone marrow.

More useful 'adult' stem cells can be harvested from the placenta once a child is born. Some companies will store your milk teeth in a special freezer so that the stem cells they contain could be used to treat you later. These stem cells can grow into a range of specialised cells.

Most adult human stem cells grow into only a few cell types, because many genes are switched off. Scientists have recently found ways to switch common adult stem cell genes back on. It might be possible to use a person's own stem cells to replace any of their damaged cells. There wouldn't be a risk of rejection and the cells might restore organs that could not otherwise heal themselves.

Some successes

The national regulatory bodies have not approved any treatments yet for widespread use, but trials and experiments are underway. Doctors need to be sure that the cells will behave in the right way when they are implanted into the body.

Claudia Castillo received a new windpipe made with cartilage produced by her own stem cells. Her lungs had been badly damaged by illness.

- A donor windpipe was treated with enzymes so that all the cells were removed.
- Stem cells were harvested from the patients' bone marrow.
- In the lab the stem cells developed into cartilage cells.
- The 'scaffolding' of connective tissue was coated with the cells.
- Surgeons implanted the new windpipe.
- The patient recovered – there was no rejection.

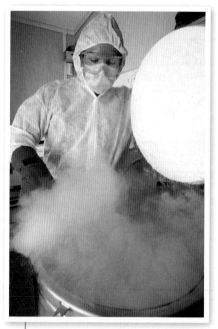

Some companies will store your baby teeth. The stem cells that they contain might help to treat you years later.

This patient received a new windpipe. The cartilage had been made using her own stem cells.

Questions

1 Why is therapeutic cloning using embryonic stem cells controversial?

2 Describe how stem cells from a donor could treat leukaemia.

3 Name two potential sources of adult stem cells.

4 What are the advantages of using stem cells from your own body to treat an illness?

5 Independent regulators oversee the use of gametes and embryos in fertility treatment and research in the UK. Why is it important that regulators are independent from government and the scientists carrying out research?

Science Explanations

Genetic technologies are at the cutting edge of modern science. The study of proteins made in cells, stem cell technology, and cellular control are key areas of research. These fields promise to provide huge benefits to present and future generations.

You should know:

- that in multicellular organisms, cells are often specialised to do a particular job
- that up to the eight-cell stage, all the cells in a human embryo are identical (embryonic stem cells) and can produce any type of human cell
- why some cells (adult stem cells) remain unspecialised, becoming specialised at a later stage
- that in plants, only cells within special regions called meristems undergo cell division – these can develop into any kind of plant cell and are used to produce clones
- about the range of tissues that unspecialised plant cells can become
- that plant hormones (auxins) cause cut stems from a plant to develop roots and grow into a complete plant, which is a clone of the parent
- that the environment affects growth and development of plants, for example, plants grow towards light (phototropism), and how this happens
- how genetic information is stored within cells
- why cell division by mitosis produces two new cells identical to each other and to the parent cell
- what happens during the main processes of the cell cycle (cell growth and mitosis)
- why a special type of cell division called meiosis is needed to produce gametes
- that instructions for making proteins are coded for by genes in the nucleus, but proteins are produced in the cell cytoplasm
- that four different bases make up both strands of the DNA molecule and that these always pair up in the same way, A with T and C with G
- that the order of bases in a gene is the genetic code to produce amino acids to make a protein
- why body cells in an organism contain the same genes, but they only produce the specific proteins they need because those genes that are not needed are switched off
- how any gene can be switched on during development in embryonic stem cells to produce any type of specialised cell
- the applications that adult stem cells and embryonic stem cells can potentially be used for.

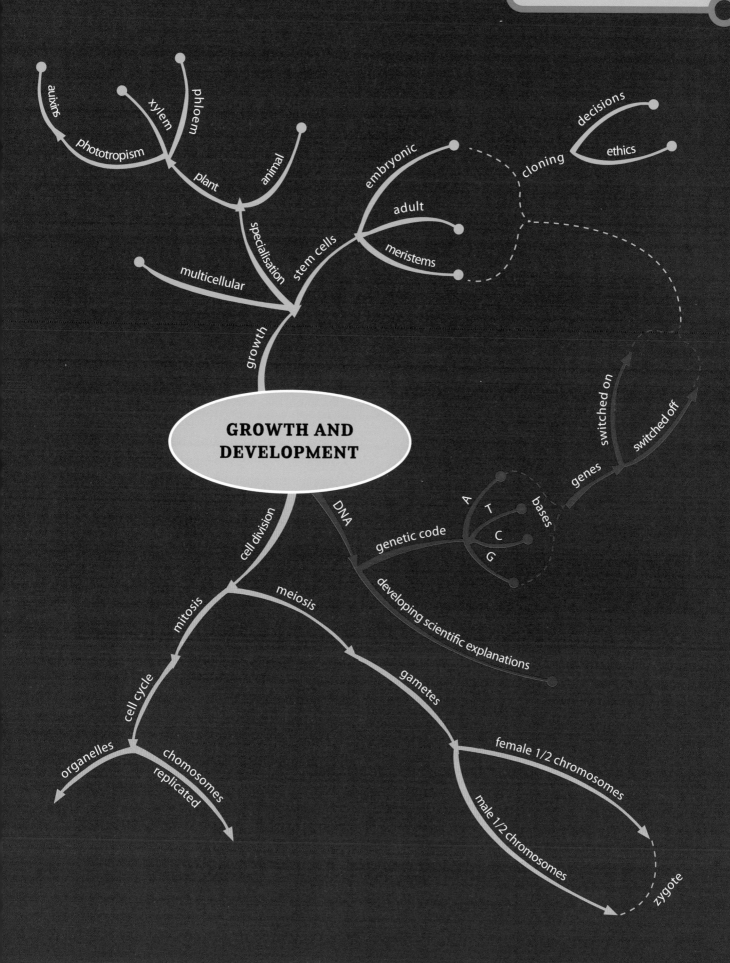

GROWTH AND DEVELOPMENT

auxins
phototropism
xylem
phloem
plant
animal
specialisation
multicellular
stem cells
embryonic
adult
meristems
cloning
decisions
ethics
growth
genes
switched on
switched off
bases
A
T
C
G
DNA
genetic code
developing scientific explanations
cell division
mitosis
meiosis
cell cycle
organelles
chomosomes
replicated
gametes
female 1/2 chromosomes
male 1/2 chromosomes
zygote

Ideas about Science

In this module you learn more about growth and development. You should become more aware that scientists need to work out explanations of their observations and measurements.

- Watson and Crick discovered the structure of DNA in 1953. They looked at all the evidence gathered in the previous 50 years and produced a model that satisfied the evidence. It showed that DNA is a double helix.

You need to distinguish between data and the explanations that explain the data. You also need to suggest whether the data and the explanation agree or not. You need to recognise when creative thought is involved in producing an explanation.

- Watson and Crick's DNA structure accounted for what was known about DNA so far. It also let them predict how DNA might copy itself exactly.

Scientific explanations allow predictions to be made. If the prediction proves to be correct, this increases the confidence in the explanation. When a prediction proves to be wrong, either the prediction or the explanation may be incorrect.

The use of stem cells has ethical implications.

- The potential of stem cells to treat diseases and mend damaged tissue is enormous. Some stem cells come from human embryos, but there is a high risk of rejection if these stem cells do not have the same genes as the person getting the transplant. There are also ethical issues about using embryos in this way.
- One solution is to replace the nucleus of a zygote with a nucleus from a body cell of the recipient. This would then develop into an embryo – a clone of the person. Stem cells from the cloned embryo could be used to treat diseases without the fear of rejection. Some people think that producing clones of people in this way is wrong. Scientists are now studying how to switch genes back on in body cells to form cells of all tissue types.
- All of this work is subject to government regulation.

It is sometimes difficult to decide what is right and what is wrong. Some people think that the right decision is one that leads to the best outcome for the greatest number of people involved.

Other people think that certain actions are either right or wrong whatever the consequences. You need to identify arguments that are based on these two different ideas.

Review Questions

1 This question is about genes.

a What is the job of genes?

 i store glucose

 ii describe how to make proteins

 iii release energy by respiration

 iv transport materials around a cell

b Complete the following statement by choosing from the table below.
Genes are sections of very ....

... long DNA molecules that make up chromosomes

... short DNA molecules that make up chromosomes

... short chromosomes that make up DNA molecules

... long chromosomes that make up DNA molecules

2 The growth and development of each cell is controlled by its DNA.
What are the features of DNA?
Make and complete a table like the one below.

DNA feature	Answer
number of strands	
number of different bases	
arrangement of bases between strands	single pairs triplets fours
shape of molecule	

3 Meiosis occurs during the formation of gametes.
Four people try to explain the link between meiosis and fertilisation.

Jo
Fertilisation causes the zygote to have the full chromosome number.

Lee
Fertilisation avoids the chromosomes mixing together inside the zygote nucleus.

Ray
Meiosis produces gametes with 23 chromosomes each.

Sue
Since the number of chromosomes in gametes is halved, twice as many gametes can be produced.

Which two people's ideas together give the best explanation of the link between meiosis and fertilisation?

4 Cells in an embryo become specialised.
What are the results of this change?
Decide whether the following statements are true or false.

The cells no longer contain the same genes.

Some of the genes are no longer active.

Each cell produces only the specific proteins it needs.

The cells form different types of tissues.

5 The order of bases in a gene is the code for building up amino acids in the correct order to make a particular protein.
Explain how the code works.

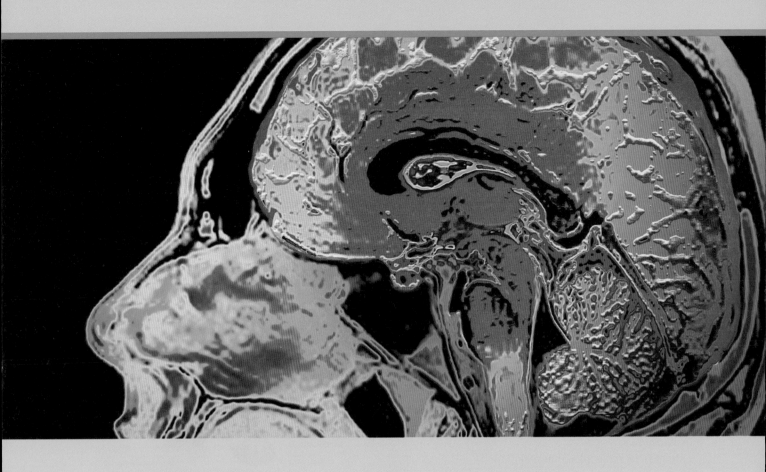

B6 Brain and mind

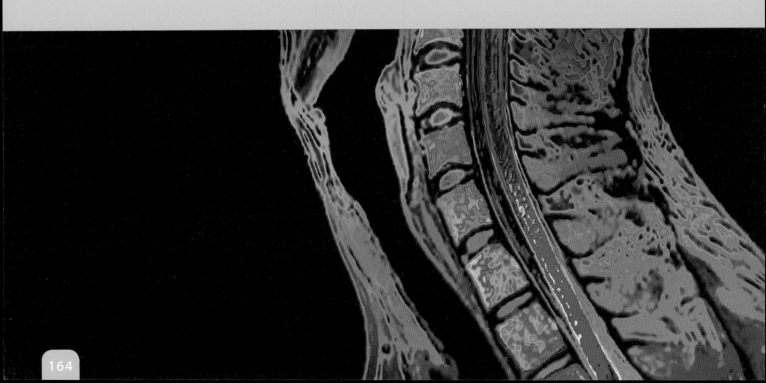

Why study the brain and mind?

The human brain allows our species to survive on Earth. It gives us advantages of intelligence and sophisticated behaviour.

What you already know

- The success of humans is mainly due to the evolution of a large and complex brain.
- Drugs can affect human behaviour.
- Nerves and hormones help you respond to your environment.

Find out about

- how organisms respond to stimuli
- how nerve impulses are passed around your body
- how your brain coordinates your senses
- how you learn new skills
- how scientists are finding out about memory.

The Science

Animals respond to stimuli in order to survive. The central nervous system – the brain and spinal cord – coordinates millions of electrical impulses every second. These impulses determine how we think, feel, and react – our behaviour. Some drugs can affect this.

Ideas about Science

Applications of scientific research can have ethical implications. Some people say that the right decision is the one leading to the best outcome for the most people. Others say that some actions are always unnatural or wrong.

Find out about

- ✔ **what behaviour is**
- ✔ **how simple behaviour helps animals survive**

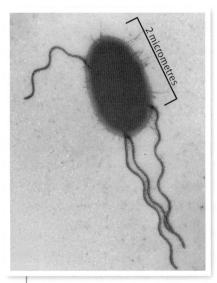

Escherichia coli bacteria are found in the lower gut of warm-blooded animals. They detect the highest concentration of food and move towards it.

Woodlice move away from light, so you are most likely to find them in dark places.

Imagine you are sitting outside. The temperature drops, and you get cold. You start to shiver.

Shivering is a **response** to the change in temperature. A change in your environment, like a drop in temperature, is called a **stimulus**. Eating is a response to the stimulus of hunger. Scratching is a response to an itch. Shivering, eating, and scratching are all examples of **behaviour**.

You can think of behaviour as anything an animal does. The way an animal responds to changes in its surroundings is important for its survival.

Simple behaviour

Simple animals always respond to a stimulus in the same way. For example, woodlice always move away from light. This is an example of a **simple reflex** response. Reflexes are always **involuntary** – they are automatic. Reflexes are important because they increase the animal's chance of survival. The photographs in this section all show reflexes.

Why are simple reflexes important?

Simple reflex behaviour helps an animal to:

- find food, shelter, or a mate
- escape from predators
- avoid harmful environments, for example, extreme temperatures.

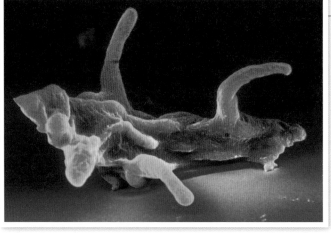

Single-celled *Amoeba* move away from high concentrations of salt, strong acids, and alkalis.

Simple reflexes usually help animals to survive. But animals that only behave with simple reflexes cannot easily change their behaviour, or learn from experience. This is a problem if conditions around them change. Their simple reflexes may no longer be helpful for survival.

When a giant octopus sees a predator, it rapidly contracts its body muscles. This squirts out a jet of water to push the octopus away from danger. The octopus may also release a dark chemical (often called 'ink'), which hides its escape.

When the tail of the sea hare *Aplysia* is pinched, the muscle contracts quickly and strongly. This reflex helps the animal escape from the spiny lobster that preys on it.

Key words
- response
- stimulus
- behaviour
- simple reflex
- involuntary

Earthworms have some of the fastest reflexes in the animal kingdom. A sharp tap from a beak on its head end is detected in the body wall. Rapid contraction of the worm's muscles pulls it back into its burrow. But this time the bird was too quick.

Have you ever tried to swat a fly? It has a very fast response to any movement that its sensitive eyes detect.

Questions

1. Write a sentence to explain each key word on this page.

2. Describe an action you did today that:
   - you did not have to think about and you have never had to learn to do
   - you have learnt to do but you can now do without thinking
   - you had to think about while you were doing it.

3. Which of the actions you described is an example of conscious behaviour, and which is most likely to be a simple reflex response?

Complex behaviour – a better chance of survival

Complex animals, like mammals, birds, and fish, have simple reflex responses. But a lot of their behaviour is far more complicated. It includes reflex responses that have been altered by experience. Also, much of their behaviour is not involuntary – they make conscious decisions. For example, if it gets very cold, you do not just rely on your reflexes to keep you warm; you decide to put on extra clothes.

Because complex animals can change their behaviour when environmental conditions change, they are more likely to survive.

Find out about

✔ **reflexes in newborn babies**
✔ **simple reflexes that help you to survive**

Behaviour in humans and other mammals is usually very complex. But simple reflexes are still important for survival. For example:

- When an object touches the back of your throat, you gag to avoid swallowing it. This is the gag reflex.
- When a bright light shines in your eye, your pupil becomes smaller. This **pupil reflex** stops bright light from damaging the sensitive cells at the back of your eye.

These types of behaviour are inherited through our genes. This is called **innate** behaviour.

Newborn reflexes

When a baby is born, the nurse checks for a set of **newborn reflexes**. Many of these reflexes are only present for a short time after birth. They are gradually replaced by behaviours learnt from experience. In a few cases these reflexes are missing at birth, or they are still present when they should have disappeared. This may mean that the baby's nervous system is not developing properly.

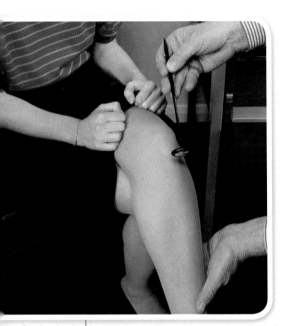

The knee-jerk reflex causes your thigh muscle to contract, so your lower leg moves upwards. Doctors may test this and other reflexes when you have a health check. Try standing still with your eyes closed. You will notice this reflex helping you to balance.

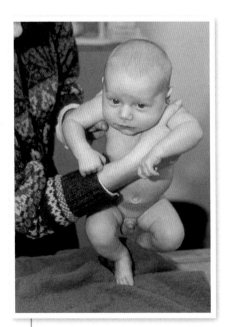

Stepping. If you hold a baby under his arms, support his head, and allow his feet to touch a flat surface, he will appear to take steps and walk. This reflex usually disappears by two–three months after birth. It then reappears as he learns to walk at around 10–15 months.

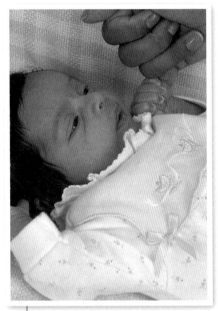

Grasping. When you put your finger in a baby's open palm, the baby grips the finger. When you pull away, the grip gets stronger. This reflex usually disappears by five–six months. If you stroke the underneath of a baby's foot, its toes and foot will curl. This reflex usually disappears by 9–12 months.

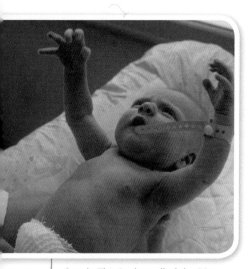

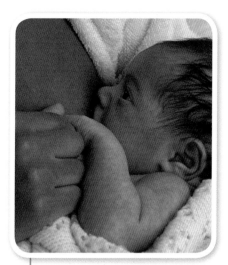

Startle. This is also called the Moro reflex, named in 1687 after the Italian scientist Artur Moro. It usually happens when a baby hears a loud noise or is moved quickly. The response includes spreading the arms and legs out and extending the neck. The baby then quickly brings her arms back together and cries. This reflex usually goes by three–six months.

Sucking. Placing a nipple (or a finger) in a baby's mouth causes the sucking reflex. It is slowly replaced by voluntary sucking at around two months.
Rooting. Stroking a baby's cheek makes her turn towards you, looking for food. This reflex helps the baby find the nipple when she is breastfeeding. The rooting reflex is gone by about four months.

Swimming. If you put a baby under six months of age in water, he moves his arms and legs while holding his breath.

Sudden infant death syndrome (SIDS)

Sudden Infant Death Syndrome, or cot death, is tragic and unsolved. In the UK about seven babies a week die from SIDS. This is 0.7 deaths for every 1000 live births.

It is likely that there are many different causes of cot death. Some people think that it could be because a baby's simple reflexes have not matured properly. This is how doctors think this may happen:

- When a fetus detects that oxygen in its blood is low, its reflex response makes it move around less. This makes sense because the less it moves around, the less oxygen it will use up in cell respiration.

- This response changes as the baby matures. When an older baby or child's airways are covered, for example, by a duvet, the baby moves more. He turns his head from side to side. He also pushes the obstruction away. So now the response to low oxygen is more activity, not less.

- If the newborn baby has not grown out of the fetal reflex, he may lie still if his bedcovers cover his airways. He is more likely to suffocate.

Doctors now advise mothers to put babies onto their backs to sleep, and not to use soft bedding like duvets. This way their faces are less likely to become covered.

Key words
- ✓ **pupil reflex**
- ✓ **newborn reflexes**

Questions

1 Describe two reflexes in:
   a adult humans
   b newborn babies.

2 How do you think the startle reflex helps a baby to survive?

3 Why are premature babies more at risk from SIDS than babies born at the correct time?

Receptors

You can only respond to a change if you can detect it. **Receptors** inside and outside your body detect stimuli, or changes in the environment.

You can detect many different stimuli, for example, sound, texture, smell, temperature, and light. Different types of receptors each detect a different type of stimulus. Receptors on the outside of your body monitor the external environment. Others monitor changes inside your body, for example, core temperature and blood sugar levels.

Sense organs

Some receptors are made up of single cells, for example, pain receptors in your skin. Other receptor cells are grouped together as part of a complex sense organ, for example, your eye. Vision is very important in humans and most other mammals. Light entering our eyes helps us humans produce a three-dimensional picture of our surroundings. This gives us information about objects such as their shape, movement, and colour.

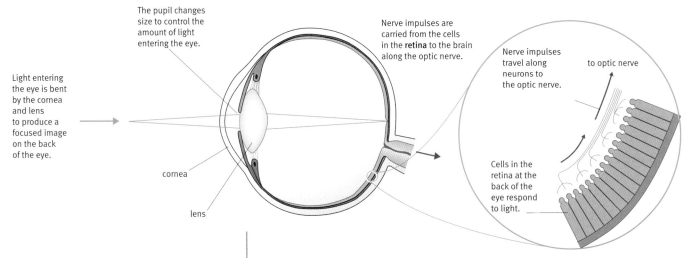

The pupil changes size to control the amount of light entering the eye.

Nerve impulses are carried from the cells in the **retina** to the brain along the optic nerve.

Nerve impulses travel along neurons to the optic nerve.

to optic nerve

Light entering the eye is bent by the cornea and lens to produce a focused image on the back of the eye.

Cells in the retina at the back of the eye respond to light.

cornea

lens

Light is focused by the cornea and lens onto light-sensitive cells at the back of the eye. These cells are receptors. They trigger nerve impulses to the brain.

Effectors

The body's responses to stimuli are carried out by **effector** organs. In multicellular organisms the effectors are either **glands** or **muscles**. Nervous and **hormonal** communication systems have developed as living things have evolved. Multicellular organisms have evolved a complex communication system that co-ordinates short-term and long-term responses.

Stimuli from the environment bring about responses in muscles or glands. Nerve impulses bring about fast, short-lived responses, for example, contractions of muscles. Hormones bring about longer-lasting effects such as an increase in growth rate.

Effectors are either glands or muscles.

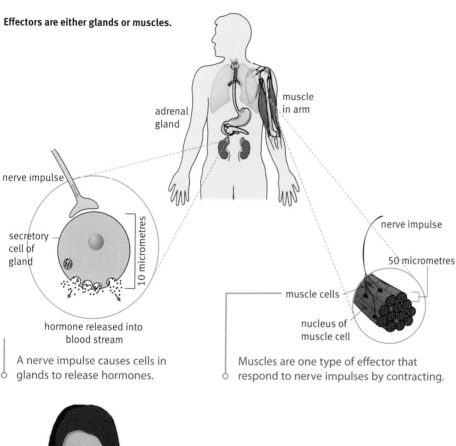

A nerve impulse causes cells in glands to release hormones.

Muscles are one type of effector that respond to nerve impulses by contracting.

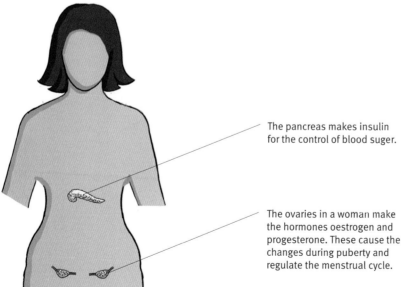

The pancreas makes insulin for the control of blood suger.

The ovaries in a woman make the hormones oestrogen and progesterone. These cause the changes during puberty and regulate the menstrual cycle.

Hormones are released when a nerve impulse reaches glands such as the ovaries and pancreas. Hormones have a slower and longer-lasting effect than a nerve impulse.

Questions

4 Name the two different types of effector and say what they do.

5 Which receptors would you use in order to thread your trainers with new laces?

6 Which effectors are you using when you:
- text a friend?
- cry?
- run a race?

7 Some people suffer from a disease where tiny clusters of light receptor cells in different parts of the eye become damaged. How would this affect what the person sees?

The process of evolution produced larger and more complex multicellular animals. These animals developed nervous and hormonal systems to allow them to respond to the environment. This gave them a survival advantage over simpler animals.

Walk out from a dark cinema on a bright afternoon and your pupils will become smaller. This pupil reflex prevents bright light damaging your eye. Like all reflexes, this behaviour is coordinated by your nervous system.

Cells in your nervous system carry **nerve impulses**. These nerve impulses allow the different parts of the nervous system to communicate with each other.

Peripheral nervous system

Many nerves link your brain and spinal cord to every other part of your body. These nerves make up the **peripheral nervous system**.

Nerves and neurons

Nerves are bundles of specialised cells called **neurons**. Like most body cells, neurons have a nucleus, a cell membrane, and cytoplasm. They are different from other cells because the cytoplasm is shaped into a very long thin extension. This is called the **axon**, and it is how neurons connect different parts of the body.

Axons carry electrical nerve impulses. Like wiring in an electrical circuit, the axons must be insulated from each other. The insulation for an axon is a **fatty sheath** wrapped around the outside of the cell. The fatty sheath increases the speed that impulses move along the axon.

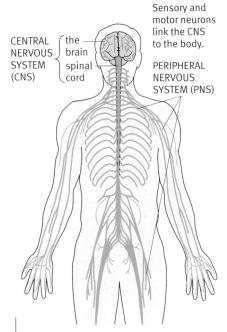

CENTRAL NERVOUS SYSTEM (CNS) { the brain, spinal cord }

Sensory and motor neurons link the CNS to the body.

PERIPHERAL NERVOUS SYSTEM (PNS)

In the mammalian nervous system the brain and spinal cord is connected to the body via the peripheral system.

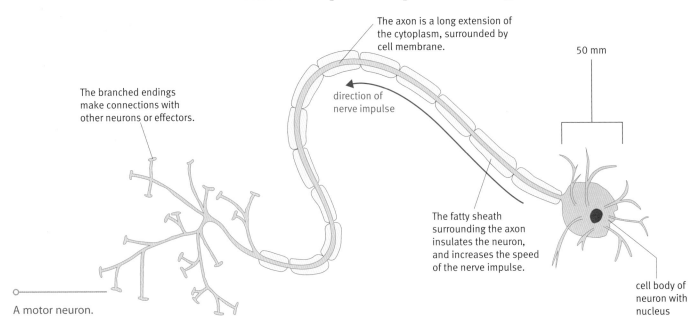

The axon is a long extension of the cytoplasm, surrounded by cell membrane.

50 mm

direction of nerve impulse

The branched endings make connections with other neurons or effectors.

The fatty sheath surrounding the axon insulates the neuron, and increases the speed of the nerve impulse.

cell body of neuron with nucleus

A motor neuron.

The reflex arc

In a simple reflex, impulses are passed from one part of the nervous system to the next in a pathway called a **reflex arc**. The diagram below shows this pathway for a pain reflex. The **relay neurons** in the spinal cord connect the **sensory neuron** to the appropriate motor neuron.

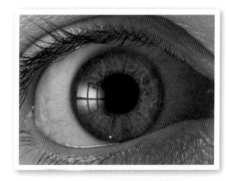

The **central nervous system** is made up of the brain and spinal cord. It coordinates the body's responses.

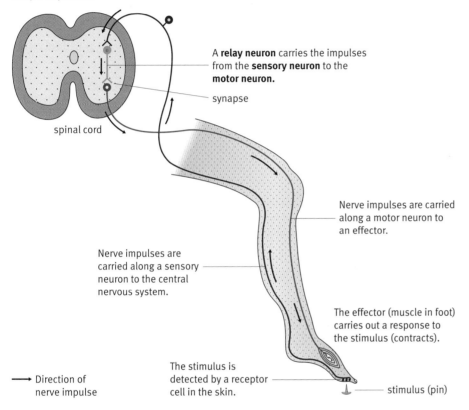

A **relay neuron** carries the impulses from the **sensory neuron** to the **motor neuron.**

synapse

spinal cord

Nerve impulses are carried along a sensory neuron to the central nervous system.

Nerve impulses are carried along a motor neuron to an effector.

The effector (muscle in foot) carries out a response to the stimulus (contracts).

→ Direction of nerve impulse

The stimulus is detected by a receptor cell in the skin.

stimulus (pin)

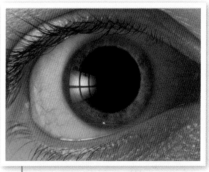

Muscles in the iris cause the pupil to change size depending on the brightness of light entering the eye. (The pupil size controls the amount of light intensity that is the stimulus.)

Central nervous system

Your **central nervous system** (CNS) coordinates all the information it receives from your receptors via sensory neurons. Information about a stimulus goes to either your brain or your spinal cord. In a reflex arc the CNS links incoming information from receptors with **motor neurons**. Motor neurons complete the reflex arc by stimulating effectors that carry out the necessary response.

Key words

- ✓ nerve impulses
- ✓ reflex arc
- ✓ central nervous system
- ✓ relay neuron
- ✓ sensory neuron
- ✓ motor neuron
- ✓ peripheral nervous system
- ✓ neurons
- ✓ axon
- ✓ fatty sheath

Questions

1 Describe the difference between:
- the job of a sensory neuron and a motor neuron
- an axon and a neuron
- the central nervous system and the peripheral nervous system.

2 Draw a labelled diagram to show a reflex arc for the newborn grasp reflex shown on page 168. This reflex is coordinated by the spinal cord.

Find out about

✓ **how nerve impulses pass from one neuron to the next**

Quick responses to stimuli are also essential in real life. They help you survive by avoiding danger.

Curare is a very powerful toxin. It is used on the tips of blowpipe darts.

Key words

✓ **synapses**
✓ **transmitter substances**
✓ **receptor molecules**
✓ **serotonin**
✓ **Ecstasy**

Think about playing a fast sport or computer game. You need very quick reactions to win. Nerve impulses give you fast reactions because they travel along axons at 400 metres per second.

Mind the gap

Neurons do not touch each other. So when nerve impulses pass from one neuron to the next, they have to cross tiny gaps. These gaps are called **synapses**. Some drugs and poisons (toxins) interfere with nerve impulses crossing a synapse. This is how they affect the human body.

How do nerve impulses cross a synapse?

Nerve impulses cannot jump across a synapse. Instead, chemicals called **transmitter substances** are used to pass an impulse from one neuron to the next. The diagram below explains how this works.

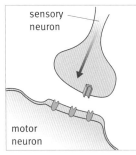
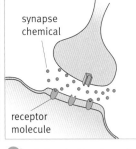
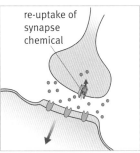

1. A nerve impulse arrives at a synapse. The direction of the impulse is shown by the arrow.

2. A chemical is released from the sensory neuron. It diffuses across the synapse.
The molecule is the correct shape to fit into **receptor molecules** on the membrane of the motor neuron.

3. A nerve impulse is stimulated in the motor neuron. The chemical is absorbed back into the sensory neuron to be used again.

How a synapse works.

Do synapses slow down nerve impulses?

The gap at a synapse is only about 20 nanometres (nm) wide. The synapse chemical travels across this gap in a very short time. Synapses do slow down nerve impulses to about 15 metres per second. A nerve impulse still travels from one part of your body to another at an incredible speed.

Being human – just chemicals in your brain?

The way we think, feel, and behave does involve a series of chemicals moving across synapses between neurons, but there is more to behaving like a human than chemicals in your brain.

These processes are very complicated. Scientists researching how your nervous system works are only just beginning to understand the brain.

Serotonin

Serotonin is a chemical released at one type of synapse in the brain. When serotonin is released, you get feelings of pleasure. Pleasure is an important response for survival. For example, eating nice-tasting food gives you a feeling of pleasure. So you are more likely to repeat eating, which is essential for survival.

Lack of serotonin in the brain is linked to depression. Depression is a very serious illness. At least one person in five will suffer from a depressive illness at some point in their life. They feel very unhappy for many days on end and often find it difficult to manage normal everyday things like working, studying, or looking after their family.

How do some drugs affect the brain?

Prozac is the name of an antidepressant drug. Prozac causes serotonin concentration to build up in synapses in the brain. So a person may feel less unhappy. The diagram explains how Prozac works. Like all drugs, Prozac can have unwanted effects.

Ecstasy

Ecstasy is the common name for the drug MDMA. Ecstasy works in a similar way to Prozac. People who have taken Ecstasy say that it can give them feelings of happiness and being very close to other people. Studies on monkeys suggest that long-term use of Ecstasy may destroy the synapses in the pleasure pathways of the brain. Permanent anxiety and depression might result, along with poor attention span and memory. For some people the harmful effects of Ecstasy are more immediate. Ecstasy interferes with the body's temperature control systems. It also slows down production of the hormone ADH in the brain. These effects can be fatal. You can read more about ADH in module B2.

Beta blockers

Some people suffer from severe chest pains called angina. This pain can be triggered when people are stressed or excited. Nerve impulses stimulate the heart to speed up. This can leave the heart muscle painfully starved of oxygen. A doctor might prescribe a drug called a beta blocker to help. Beta blockers reduce the transmission of impulses across nerve synapses, stopping the heart beat from speeding up. Beta blockers also help to control nerve impulses inside the heart, making sure that the heart beats in a regular, controlled way.

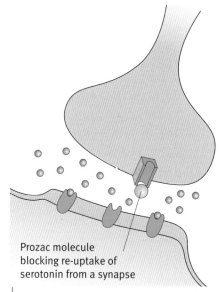

Prozac molecule blocking re-uptake of serotonin from a synapse

Feelings of depression can be caused by too little serotonin in the brain. Prozac works by blocking re-uptake of serotonin.

Questions

1 Write down a sentence to describe a synapse.

2 Draw a flow diagram to describe what happens when a nerve impulse arrives at a synapse.

3 Explain how the release of serotonin in the brain helps us survive.

4 Some drugs (like curare) block the receptors on the motor neuron at a synapse. Explain how this would affect muscles that the motor neuron is linked to.

5 Prozac is described as a selective serotonin re-uptake inhibitor or SSRI. Explain how an SSRI might help a person with depression.

Find out about

- ✓ **the structure of your brain**
- ✓ **how scientists learn about the brain**

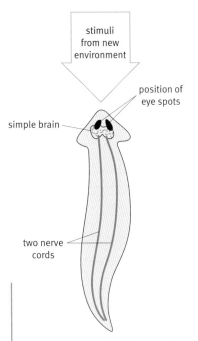

Sense organs on the flatworm head detect light and chemical stimuli. A simple brain processes the response.

Think of some things you did today. Getting up, deciding what to have for breakfast, travelling to school, talking, and listening to friends. All of this complex behaviour has been controlled by your brain. But exactly how it happens is still being researched. Scientists who study the brain are called **neuroscientists**. Neuroscience is a fairly 'new' science. This means that scientists have only recently started to investigate how the brain works.

Complex nervous and hormonal communication systems only developed once multicelllular organisms evolved. These organisms have specialised tissues and organs to carry out communication processes.

Simple animals

Neurons carry electrical impulses around your body. Quite simple animals have a larger mass of neurons at one end of their body – the head end. This end reaches new places first, as the animal moves. These neurons act as a simple brain. They process information coming from the receptors on their head end.

Looking at how the brains of simple animals work can help scientists begin to understand more complicated brains.

Complex animals

More complex behaviour like yours needs a much larger brain. So your brain is made of billions of neurons. It also has many areas, each carrying out one or more specific functions all in the same organ. Your complex brain allows you to learn from experience, for example, how to behave with other people.

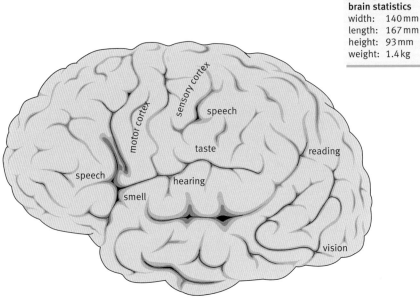

Average human brain statistics
width: 140 mm
length: 167 mm
height: 93 mm
weight: 1.4 kg

This is a diagram called the 'sensory homunculus'. Each body part is drawn so that its size represents the surface area of the sensory cortex that receives nerve impulses from it.

The human cerebral cortex is a highly folded region. Although it is only 5 mm thick, its total area is about 0.5 m². This map of the cerebral cortex shows the regions responsible for some of its functions.

The conscious mind

When you are awake, you are aware of yourself and your surroundings. This is called **consciousness**. The part of your brain where this happens is the **cerebral cortex**. This part is also responsible for intelligence, language, and memory. Brain processes to do with thoughts and feelings happen in the cerebral cortex and are what is called your 'mind'.

The cerebral cortex is very large in humans compared with other mammals. Studying what goes on in this part of the brain helps us to understand what it means to be human.

Finding out about the brain

In the 1940s a Canadian brain surgeon, Wilder Penfield, was working with patients who had epilepsy. Penfield applied electricity to the surface of their brains in order to find the problem areas. The patients were awake during the operations. There are no pain receptors in the brain so they did not feel pain.

Penfield watched for any movement the patient made as he stimulated different brain regions. From this information he was able to identify which muscles were controlled by specific regions of the motor cortex.

Injured brains

Scientists study patients whose brains are partly destroyed by injuries or diseases like strokes. Studies of injured soldiers have been important for research on how the brain functions.

Brain imaging

Modern imaging techniques such as magnetic resonance imaging (MRI) scans provide detailed information about brain structure and function without having to open up the skull. MRI can be used to show which parts of the brain are most active when a patient does different tasks. These scans are called functional MRI (fMRI) scans. The active parts of the brain have a greater flow of blood.

Key words

✔ **consciousness**
✔ **cerebral cortex**
✔ **neuroscientist**

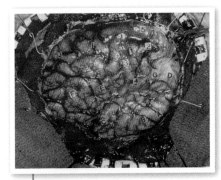

Penfield mapped the motor cortex by stimulating the exposed brain during open brain surgery. Regions of this brain have been identified and labelled in a similar way.

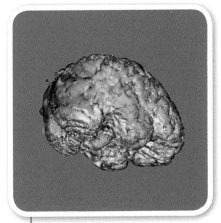

This functional MRI scan shows up areas of the brain that are active as a patient carries out a specific task. This patient was reading aloud.

Questions

1 What is your brain made up of?

2 Why is a complex brain so important for survival?

3 Which four functions of the brain happen in the cerebral cortex?

4 Explain why it is necessary for blood flow to increase to parts of the brain that are very active.

5 Compare the diagram of the brain with the functional MRI scan. How can you tell that the person was reading aloud?

6 What ethical issues should scientists consider when using injured humans to study the brain?

Find out about

✓ **how conditioned reflexes can help you survive**

Pavlov's experiment

Pavlov's dog salivated when presented with food.

The food is the stimulus and salivation is the response.

Pavlov rang a bell while his dog was eating its food.

After a while the dog salivated when it heard the bell, even if no food was around.

The dog had learned to link the stimulus of ringing the bell with food. This type of learning is called **conditioning**.

Learning to link a new stimulus with a reflex action allows animals to change their behaviour. This is called a **conditioned reflex**. The final response – salivation – has no direct connection to the primary stimulus – food.

The lion cub below is just a few weeks old. She was born with reflexes that are helping her to stay alive. But much of her behaviour, for example, how to hunt or how to get on with other lions in the pride, she will **learn** from her mother. This is learnt behaviour.

Learnt behaviour is just as important for this lion cub's survival as reflexes.

Being able to learn new behaviour by experience is very important for survival. It means that animals can change their behaviour if their environment changes.

Conditioned reflexes

In 1904 the Russian scientist Ivan Pavlov won a Nobel Prize for his study on how the digestive system works. In his research Pavlov trained a dog to expect food whenever it heard a bell ring. The diagrams on the left explain what happened.

Conditioning aids survival

Conditioned reflexes can help animals survive. For example, bitter-tasting caterpillars are usually brightly coloured. A bird that tries to eat one learns that these bright colours mean that caterpillars will have a nasty taste. After a first experience the bird responds to the colours by leaving them alone. So this helps the caterpillars to survive.

If the brightly coloured insect is also poisonous, this reflex will help the bird survive as well. If other very tasty insects have similar colours and patterns, the bird does not eat them because of this conditioned response. You might have been caught out by this too – harmless yellow-and-black striped hoverflies sometimes alarm people who have been stung by a wasp.

'Warning' colours protect this caterpillar from predators.

Conditioning your pet

Open a can of soup in your kitchen. If you have a dog or a cat, this sound may get them very interested. But they are not hoping for soup! The animal's reflex response to food has been conditioned. It has learnt through experience that the sound of a tin being opened may be followed by food being put into its dish.

If a cat only uses its basket when you are taking it to the vet for an injection, it may become conditioned to link the basket with a frightening experience. The cat will then always be frightened by the stimulus of the basket. It will fight to keep out of the basket, even if you are only trying to take it to a new home.

Goldfish become conditioned to expect food when they see you in the room. They swim to the front of the bowl when you appear. The goldfish are linking the stimulus of seeing you with the original stimulus – food in the water.

Questions

1   Draw a flow diagram to explain how a cat or other pet can become conditioned to expect food when it hears a bathroom shower being run.

Use the key words from this section in your answer.

2   Adverts often have glamorous, funny, or exciting images and catchy tunes. Write down a list of photos and tunes from adverts that remind you of things you could buy. How is conditioning involved in making us more likely to buy these products?

Key words
- ✓ **learn**
- ✓ **conditioning**
- ✓ **conditioned reflex**

Hundreds of neurons interact to coordinate the responses you make when you are receiving this many stimuli.

Hundreds of complex pathways in the brain have to be used to succeed in this fast-moving sport.

This complicated behaviour involves highly complex pathways in the brains of both these animals.

Conscious control of reflexes

Most human reflex arcs are coordinated by the spinal cord. Only reflexes with receptors on the head are coordinated by the brain. A reflex arc only has simple connections between a sensory neuron and a motor neuron.

You do not think about reflex responses – your brain does not have to make a decision. They happen automatically, because they are designed to help you survive.

But sometimes a reflex may not be what you want to happen. Some reflexes can be modified by **conscious** control. Imagine picking up a hot plate. Your pain reflex makes you drop it. But if your dinner is on the plate, you can overcome this reflex and hold onto the plate until you put it down safely. The conscious control of your brain overcomes the reflex response. The diagram below explains how this happens.

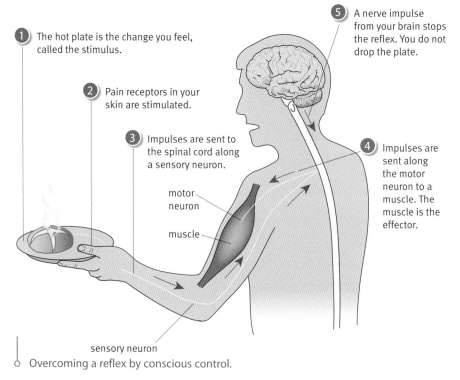

1. The hot plate is the change you feel, called the stimulus.

2. Pain receptors in your skin are stimulated.

3. Impulses are sent to the spinal cord along a sensory neuron.

motor neuron

muscle

5. A nerve impulse from your brain stops the reflex. You do not drop the plate.

4. Impulses are sent along the motor neuron to a muscle. The muscle is the effector.

sensory neuron

Overcoming a reflex by conscious control.

It's all in the mind – more complex behaviour

Connections in the brain usually involve hundreds of other neurons with different connections. Using these complex pathways, your brain can process highly complicated information, such as music, smells, and moving pictures. Different parts of the brain also store information (memory) and use it to make decisions for more complicated behaviour.

Complex behaviour allows us to learn from experience. For instance, **social behaviour** is learnt as humans develop.

Early humans learnt how to make and use tools for food and protection. Their ability to learn language meant that they could communicate new ideas. This gave them a survival advantage.

Making ethical decisions

Our understanding of the brain and behaviour has mainly come from experimenting on animals and (in the past) humans. This has allowed us to improve our theories about human learning, and to develop new treatments for diseases and injuries.

Some people argue that using animals for medical research is acceptable. Other people think that tampering with a vertebrate's brain in the name of science cannot be justified.

Scientists have learnt a lot about the brain from studying humans with mental-health problems. But is it fair and right to experiment on epileptic patients, as Penfold did? Some people argue that experimenting on ill patients results in improved medical knowledge that will benefit many others.

Studies of soldiers whose brains were damaged in war have helped scientists learn how the human brain works. But new technologies like MRI now mean that it is possible to build up a clearer picture of how the brain works just by observing a healthy brain in action.

> ### Key words
> - ✓ conscious
> - ✓ social behaviour

Questions

3 List two reflexes you can overcome, and two that you cannot.

4 A baby urinates whenever its bladder is full. Draw a labelled diagram to show how nerve impulses from the brain overcome this reflex when the child is older.

5 Give three examples of how early humans' ability to learn gave them a survival advantage.

6 Do you think there is any ethical difference between using a rat and using a monkey for experiments to find out how the brain works?

7 Under what circumstances do you think it could be right to conduct scientific experiments on a human with a brain disorder?

Find out about

- ✔ **how human beings learn new things**
- ✔ **explanations that scientists have for how your memory works**

Mammals have complex brains made up of millions of neurons. When humans and other mammals experience something new, they can develop new ways of responding. Experience changes human behaviour, and this is called learning. The way that people and animals behave towards each other socially is also learned.

The evolution of larger brains gave some early humans a better chance of survival. Intelligence, memory, consciousness, and language are complex functions carried out by the outer layer of the brain, which is called the cerebral cortex. These functions are all involved in learning.

How does learning happen?

Neurons in your brain are connected together to form complicated **pathways**. How do these pathways develop? The first time a nerve impulse travels along a particular pathway, from one neuron to another, new connections are made between the neurons. New experiences set up new neuron pathways in your brain.

If the experience is repeated, or the stimulus is particularly strong, more nerve impulses follow the same nerve pathway. Each time this happens, the connections between these neurons are strengthened. Strengthened connections make it easier for nerve impulses to travel along a pathway. As a result, the response you produce becomes easier to make.

The brains of human babies develop new nerve pathways very quickly. Your brain can develop new pathways all your life. This means you can still learn as you get older, though more slowly.

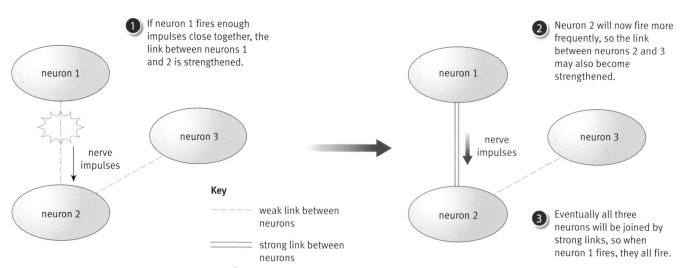

1. If neuron 1 fires enough impulses close together, the link between neurons 1 and 2 is strengthened.

2. Neuron 2 will now fire more frequently, so the link between neurons 2 and 3 may also become strengthened.

Key

– – – weak link between neurons

=== strong link between neurons

3. Eventually all three neurons will be joined by strong links, so when neuron 1 fires, they all fire.

Nerve pathways form in a baby's brain as a result of a stimulus from its environment. Repeating the stimulus strengthens the pathway. The baby then responds in the same way each time it receives the stimulus. Some neurons in the brain do not take part in any pathway. Many of these unused neurons are destroyed.

Repetition

Repetition helps you learn because it strengthens the pathways the brain uses to carry out a particular skill. Perhaps you have learned to ride a bicycle, play a musical instrument, perform a new dance sequence, or touch type. To do these things you created new nerve pathways then strengthened them through repetition. This made it easy for you to respond in the way that you practised.

For example, Marie is a gymnast. When she has to learn new movements she stands still and imagines going through the motions – the position of her body and muscles being used at each stage. Visualisation works because thinking about using a muscle triggers nerve impulses to that muscle. This strengthens the pathways the impulse takes. After a period of visualisation, the actual movement is a lot easier to perform.

Marie visualises new movements to help her learn them.

Age and learning

You learn to speak through repetition because you are surrounded by people talking. Children learn language extremely easily up to the age of about eight years. Their brains easily make new neuron pathways in the language processing region. As we get older it becomes harder for this part of the brain to make new pathways.

Feral children

In 1799, in southern France, a remarkable creature crept out of the forest. He acted like an animal but looked human. He could not talk. The food he liked and the scars on his body showed he had lived wild for most of his life. He was a wild, or **feral**, child. The local people guessed that he was about twelve years old and named him Victor.

Victor was taken to Paris. He lived with a doctor who tried to tame him and teach him language. At first people thought Victor had something wrong with his tongue or voice box. He could only hiss when people tried to teach him the names of objects. He communicated in howls and grunts.

Victor never learnt to say more than a couple of words. By the time he was found, the time in his development when it was easy to learn language had passed.

Key words
- **pathways**
- **repetition**
- **feral**

Questions

1 Write a few sentences to explain how you learn by experience. Use the key words 'pathway' and 'repetition' in your answer.

2 Explain why repeating a skill helps you learn it.

3 Write a list of skills you could practise by visualisation.

Psychologists are scientists who study the human mind. They describe **memory** as your ability to store and retrieve information.

Short-term and long-term memory

Read this sentence:

- As you read this sentence you are using your **short-term memory**.

Short-term memory lasts for about 30 seconds in most people. If you have no short-term memory you will not be able to make sense of this sentence. By the time you get to the end of the sentence, you will have forgotten the beginning.

Think about a song you know the words to:

- To remember the words you use your **long-term memory**.

Verbal memory is *any information* you store about words and language. It can be divided into short-term and long-term memory. Long-term memory is a lasting store of information. There seems to be no limit to how much can be stored in long-term memory. And the stored information can last a lifetime.

Different memory stores work separately

People with advanced **Alzheimer's disease** suffer severe short-term memory loss. They cannot remember what day it is, or follow simple instructions. But they may still remember their childhood clearly.

Some people lose long-term memory because of brain damage or disease. Their short-term memory is normal. This evidence is important because it shows that long-term and short-term memory must work separately in the brain.

The 'Nun Study' at the University of Kentucky, USA, has had the participation of 678 School Sisters of Notre Dame. They range in age from 75 to 106 years. The sisters have allowed scientists to assess their mental and physical function every year and to examine their brains at death. The study has led to significant advances and discoveries in the area of Alzheimer's disease and other brain disorders.

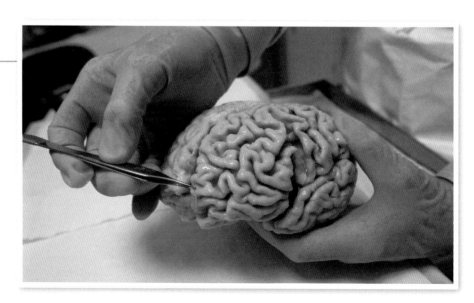

Sensory memory store

You can also use a sensory memory store to store sound and visual information for a short time. When you wave a sparkler on bonfire night it leaves a trail of light. You can even write shapes in the air that other people can see. You see the trail because you store each image of the sparkler separately for a short time. The ability to store images for a short time makes the separate pictures in a film seem continuous. You can store sound temporarily in the same way.

Your sensory memory stores each image of the sparkler separately for a short time. This makes the whole shape seem continuous.

Questions

1 Write down one sentence to describe memory.

2 What is the difference between short-term and long-term memory?

3 Explain why a person with advanced Alzheimer's disease is unable to do simple things like go shopping or cook for themselves.

4 Give one piece of evidence that short-term and long-term memory are separate.

Key words
- ✔ memory
- ✔ short-term memory
- ✔ long-term memory
- ✔ Alzheimer's disease

Key words

✔ retrieval of information
✔ models of memory
✔ multistore model

How much can you store in your short-term memory?

Cover up the list of letters below with a piece of paper. Move the paper down so you can see just the top row. Read through the row once, then cover it up and try to write down the letter sequence. Then go down to the next row and do the same. Find out how many letters you can remember in the correct order.

```
N T
A N L
N F E K
B F E X A
N A Z T P L
M B T F E Q P
U N D A C X Z G
O R B V E X Z D A
R T L D C A G P V E
```

If you remembered more than seven letters in a row correctly, then you have excellent short-term memory. Short-term memory can only store about seven items. When you are remembering letters in a list, each letter is an 'item'. To remember more letters, chunk them into groups.

For example, the row O R B V E X Z D A has nine letters. Chunk these into groups of three: 'ORB' 'VEX' 'ZDA'.

The nine letters are easy to remember because now they are only three items. Three items doesn't overload your short-term memory.

Models of memory

The **retrieval of information**, such as word lists, is a way of testing your memory. Memory tests can tell us what memory can and cannot do. But they do not explain how the neurons in the brain work to give you memory. Explanations for how memory happens are called **models of memory**.

The multistore model: memory stores work together

Read through the list of words below once. Then cover the page and try to write down as many of the words as you can remember. They can be in any order.

> dog, window, film, menu, archer, slave, lamp, coat, bottle, paper, kettle, stage, fairy, hobby, package

How many did you remember? If this type of test is carried out on large numbers of people, a pattern is seen in the words they recall. People often remember the last few words on the list and get more of them right. They also recall the first few words on the list quite well.

When you look at a list of words:

* Nerve impulses travel from your eyes to your sensory memory.
* Some sensory information is passed on to your short-term memory. Only the information you pay attention to is passed on. You will not be able to remember words you have not noticed.
* If more information arrives than the short-term memory can hold then some is lost (forgotten). You will not remember these words either.
* Some information is passed to your long-term memory. These are words you will remember – usually the first few words on the list.
* The last information your short-term memory receives will still be there when you start to write down the list. So these are also words you will remember, usually the last few words on the list.

This use of sensory, short-term and long-term memory is known as the **multistore model** of memory.

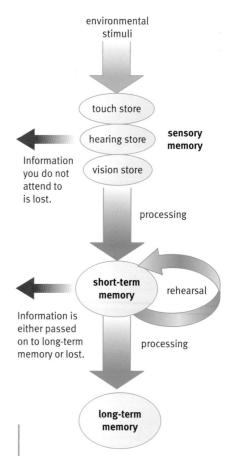

The multistore model of memory can be used to explain how some information is passed to the long-term memory store and some information is lost.

Questions

5 You read the menu on a board inside a café. When you try to tell your friend sitting outside all the choices, you forget some. Why can you not remember everything on the list?

6 'So far none of the models of memory can explain completely how your memory works.'
   a What does it take for an explanation to be widely accepted by scientists?
   b Why is it useful to have models that try to explain how memory works?

Rehearsal is one technique actors use to learn their lines.

Rehearsal and long-term memory

Look at this row of letters:

R T I D A C G P E V

There are too many letters in this row for you to store them separately in your short-term memory. Given time you would probably repeat the letters over and over until you remembered them. **Repetition of information** is a well-known way of memorising things. An actor can memorise a sonnet (a 14-line poem) in around 45 minutes. Psychologists think that rehearsal moves information from your short-term memory to your long-term memory store.

The working-memory model

In 1972, two psychologists, Fergus Craik and Robert Lockhart, concluded that the multistore memory model was too simple. They suggested that rehearsal is only one way to transfer information from short-term to long-term memory.

Rehearsed information is processed and stored rather than lost from short-term memory. Craik and Lockhart argued that you are more likely to remember information if you process it more deeply. They suggested that this will happen if you understand the information or it means something to you.

For example, if you can see a pattern in the information, you process it more deeply. So

AAT, BAT, CAT, DAT, EAT

is much easier to remember than

DAT, AAT, EAT, CAT, BAT

You also process information more deeply if there is a strong stimulus linked to the information, for example, colour, light, smell, or sound.

An active working memory

Short-term memory is now seen as an active '**working memory**'. Here you can hold and process information that you are consciously thinking about. Communication between long-term and working memory is in both directions. This way you can retrieve information you need, and also store information you may need later.

Putting it into practice

You can apply what the psychologists have discovered to your own school work.

- *Repetition:* If you are struggling to remember a piece of information you have read, read it several times.
- *Rehearsal:* Read sections of what you have to learn that are short enough to keep in your short-term memory. Make notes from memory to help move the information to your long-term memory.
- *Active memory:* Use highlighter pens and spider diagrams to process information for learning.

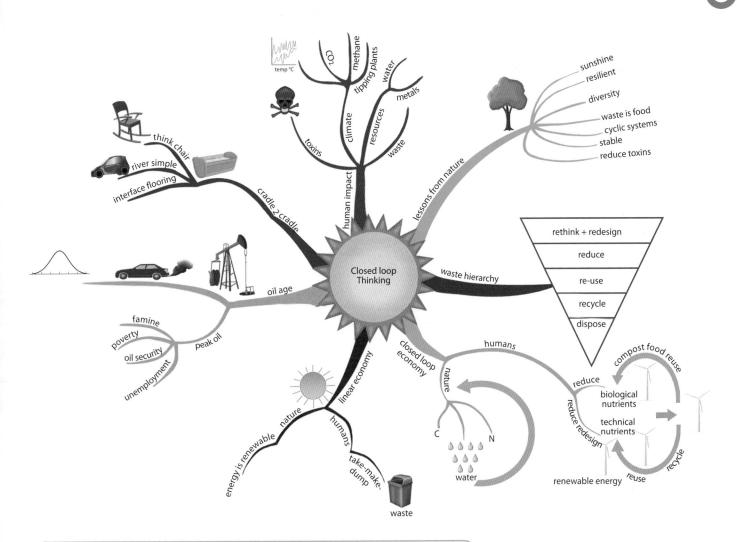

Questions

7 Make two lists of 10 different things to buy from a supermarket. Try to remember one list. Put the second list into 'families', for example tins, bakery, cleaning, and memorise it. Which list is easier to remember and why?

8 Give an example of something you can remember because a strong stimulus is linked to the information:
   a a colour   b a smell   c a sound

9 Write down an example of something you have memorised through rehearsal, for example, the directions to the cinema, or a complicated set of moves in a computer game. How has rehearsal helped you to remember this?

10 Explain why using a highlighter pen to pick out key facts makes your revision more successful.

11 Construct a spider diagram to show the information you have learnt about memory and learning.

Key words
- **repetition of information**
- **working memory**

Science Explanations

There have been big advances in neuroscience recently, but we still do not know everything about how the human brain works. Knowing how the brain works is very important for understanding how we learn and about the mental-health diseases of old age.

You should know:

- that a stimulus is a change in the environment of an organism
- what simple reflexes are and why they are important in the role of receptors, processing centres, and effectors in body systems such as human vision
- that the body uses electrical impulses and chemical hormones for short- and long-lived responses, respectively
- the relationship between the central nervous system and peripheral nervous system (sensory and motor neurons) in humans and other vertebrates
- that transmitter substances carry nerve impulses from one nerve to the next at the synapse and how some toxins and drugs affect the transmission
- that nerve impulses travel through relay neurons in the spinal cord and connect sensory and motor neurons, allowing automatic, rapid responses
- how scientists map the regions of the cerebral cortex to particular functions
- how the evolution of a larger brain gave some early humans a better chance of survival
- the role of short-term memory and long-term memory in the storage and retrieval of information
- what helps humans to learn and recall information
- how the 'multistore model' of memory provides a working model for short-term memory, long-term memory, repetition, storage, retrieval, and forgetting
- how simple models like the multistore model develop into more complex models such as the working-memory model.

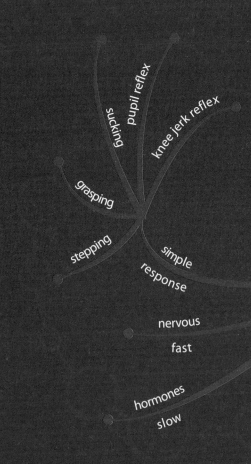

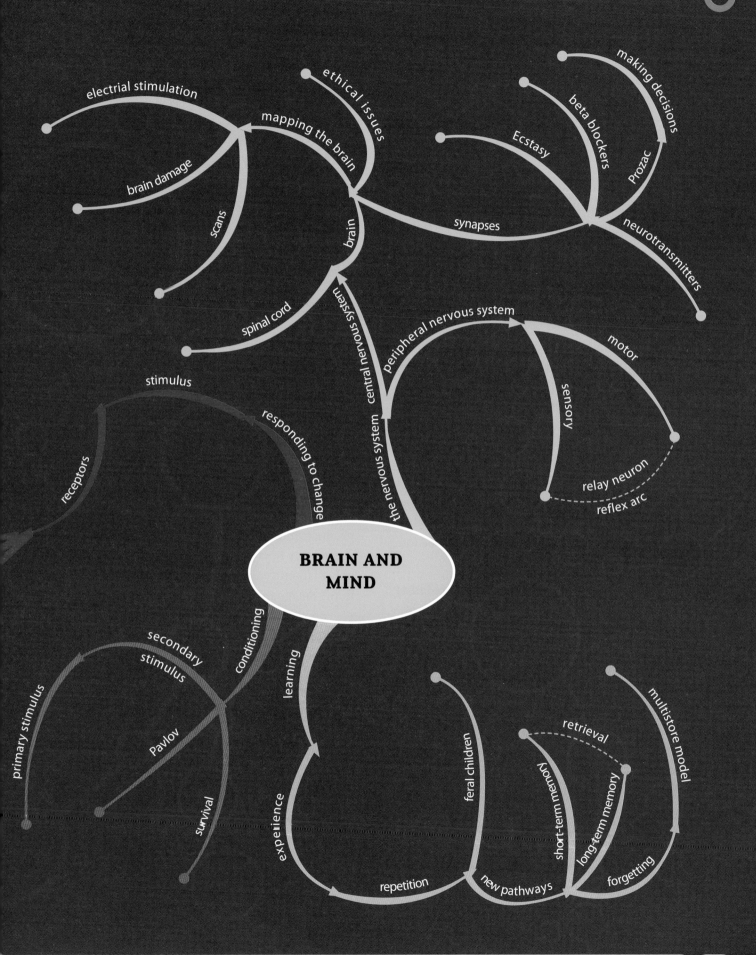

BRAIN AND MIND

electrial stimulation

ethical issues

mapping the brain

brain damage

scans

brain

spinal cord

central nervous system

the nervous system

peripheral nervous system

synapses

Ecstasy

beta blockers

making decisions

Prozac

neurotransmitters

motor

sensory

relay neuron

reflex arc

stimulus

receptors

responding to change

conditioning

secondary stimulus

primary stimulus

Pavlov

survival

learning

experience

repetition

new pathways

feral children

short-term memory

long-term memory

retrieval

multistore model

forgetting

Ideas about Science

Most knowledge about the brain has come from experiments on animals and humans. This has allowed scientists to refine theories about human learning and to develop new treatments for diseases and injuries.

You need to identify and develop arguments around ethical issues in scientific work, and summarise the different views that may be held.

- Some people argue that using animals for medical research is acceptable if there are benefits for humans. Other people think that tampering with a vertebrate's brain cannot be justified.
- Scientists have learnt a lot about the brain from studying humans with mental-health problems. But is it fair and right to experiment on people who have an illness?
- Studies of soldiers whose brains were damaged in war have helped scientists learn how the human brain works.

It is sometimes difficult to decide what is right and what is wrong. Some people think that the right decision is one that leads to the best outcome for the greatest number of people.

Other people think that certain actions are either right or wrong whatever the consequences. You need to be able to identify and develop arguments that are based on these two different ideas.

- With new technologies such as MRI it is possible to build up a picture of how the brain works by observing a healthy brain in action. This reduces the need for controversial experiments on animals and ill people.

This module examines some models that were developed to explain how humans remember and learn things.

- The simple multistore model is useful for only some of the observations or data on learning and memory. Fergus Craik and Robert Lockhart concluded in 1972 that the multistore memory model was too simple.

Craik and Lockhart used creative thought to produce another explanation to explain more completely how people remember. You should be able to identify where creative thinking is involved in the development of an explanation.

- The working-memory model provides a way of explaining a wider range of how people remember things.

When looking at science explanations you need to identify the better of two given scientific explanations for a phenomenon, and give reasons for your choice.

Review Questions

1 The gaps between sensory and motor neurons are called synapses. When an impulse is transmitted a series of events take place at the synapse.
These statements are in the wrong order.
One statement is incorrect.

A Chemicals are released into the synapse.
B The receptor molecules produce chemicals.
C Chemicals bind with receptor molecules on the motor neuron membrane.
D Chemicals diffuse across the synapse.
E The impulse travels along a motor neuron.
F An impulse reaches the end of a sensory neuron.

Select the five correct statements and write them in the correct order.

2 Pavlov used salivation in dogs to study conditioned reflexes. Pavlov used a series of steps over a period of time to produce a conditioned reflex.

a At each step a **different** stimulus was provided.
  A Dog hears bell ringing.
  B Dog shown food.
  C Dog shown food and hears bell ringing.

Copy and complete the table below by writing A, B, or C in the blank cells.

step 1: initial reflex		
	dog salivates	dog given food

step 2: repeated many times		
	dog salivates	dog given food

step 3: conditioned reflex		
	dog salivates	dog given food

b Conclusions can be made following this investigation.

Decide which of these conclusions are true and which are false.

The bell was used as a primary stimulus.

The conditioned reflex response had a direct connection to the primary stimulus.

The dog learned to associate the secondary stimulus with the primary stimulus.

3 Pip is a young puppy. His brain contains billions of neurons.
Explain what will happen to the neuron pathways in Pip's brain as he develops.

4 Scientists can gather useful information about how the brain works by studying people with brain damage.
Write down the ethical issues that are involved with this research and two different views that may be held.

5 Explain how recreational drugs such as Ecstasy can affect the transmission of impulses across the synapses between neurons in the brain.

6 Describe one model that can be used to understand how the brain stores and retrieves information as memories.

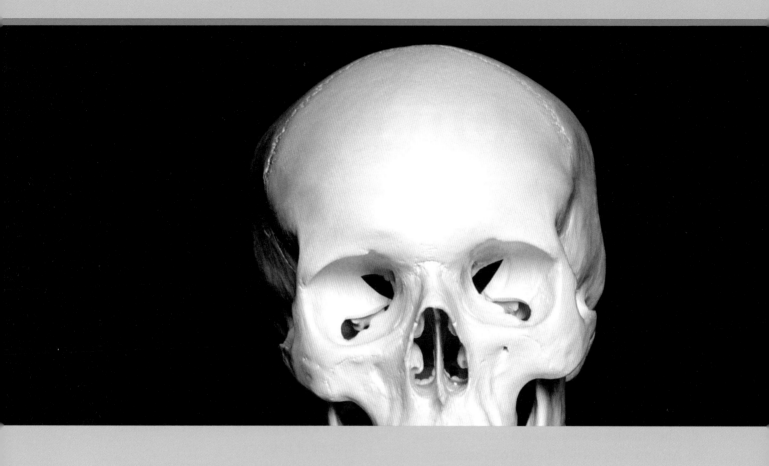

B7 Further biology

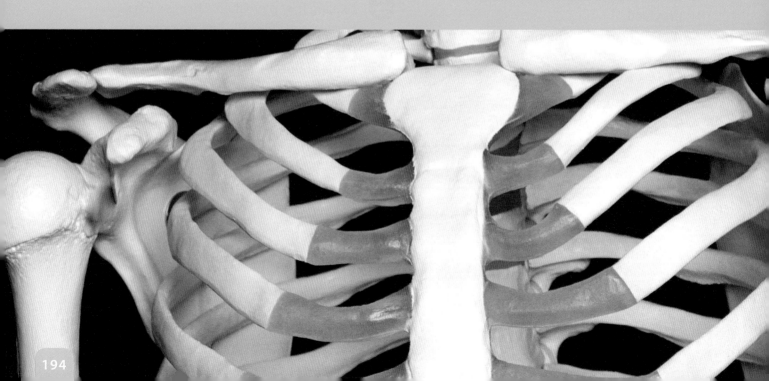

Why study further biology?

Biology is the science of life, and biological knowledge and understanding will be increasingly important for humans and their relationship with the natural world during this century.

What you already know

Further biology builds on your knowledge about the human body, how ecosystems function and new technologies.

Find out about

- how our skeleton, joints, and muscles support our body and allow us to move

- how we can monitor our level of fitness

- how our heart and blood function as a transport system

- how we keep our body temperature balanced

- how exercise and a healthy diet can help prevent illnesses like diabetes

- how natural ecosystems provide us with food, water, timber, clean air, and shelter

- the difference between a linear lifestyle and a closed-loop lifestyle

- the way materials cycle around natural ecosystems

- solutions for fishing and timber harvesting to make them sustainable

- better ways to live in a post-oil world

- using bacteria and fungi to produce medicines and enzymes

- genetic modification of bacteria and plants

- DNA technology, nanotechnology, stem cell technology, and biomedical engineering.

The Science

In 'Peak performance' you will learn more about how human bodies work and how to keep fit and healthy. In 'Learning from natural ecosystems' you will find out how the natural world provides a habitat for humans and a model for sustainable systems. In 'New technologies' you will discover more about modern biological techniques, with implications for human food and medicine production.

Ideas about Science

Data collection and analysis are important in all areas of biology. The validity of conclusions from data needs evaluation. Thinking about cause and effect is important too, and it is often useful to develop scientific explanations. We need to think if new technologies are ethically right. Risks need to be taken into account too.

Topics B7.1–3: Peak performance

Why study peak performance?

Manufactured devices can only begin to imitate the incredible endeavours that every one of us can perform. Human performance is a triumph of intelligence, sense, and control.

To keep at your peak performance it helps if you understand how your body works. Striving for peak performance has positive impacts on your health and well being.

The Science

As vertebrates, we have a jointed internal skeleton, which is moved by muscles that work in pairs. A physiotherapist can help if we have an injury.

We can monitor our levels of fitness accurately using a range of techniques. A body mass index calculation based on our height and mass can quickly tell us if we need to make changes to our diet and lifestyle.

Our heart is a double pump – its muscular walls and valves circulate blood around the lungs and the body. Blood supplies every part of our body with nutrients and oxygen whilst removing waste products.

The conditions inside our body need to remain constant for our cells to function properly. If conditions change from their normal level then our body responds until the balance is restored. If we choose a poor diet and don't exercise, we become vulnerable to illnesses such as diabetes, heart disease, and some cancers.

Ideas about Science

Measurements of performance rely on collecting repeatable and accurate data.

Data on health and performance from a human population lies within a range due to variation between people, and other factors.

Deciding what is 'average' or 'normal' needs to refer to mean values, the spread of the data, and any outliers in the data.

Scientists must be aware that when two factors are correlated, this does not mean one causes the other.

Find out about

✓ **the skeletal system**

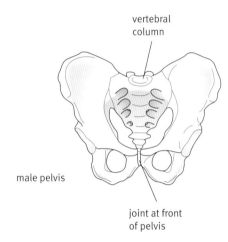

vertebral column

male pelvis

joint at front of pelvis

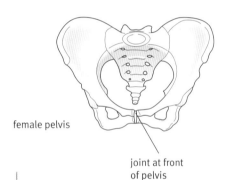

female pelvis

joint at front of pelvis

The adult human male and female pelvic girdles look different. The female's pelvis is shallower and wider for childbearing. This fact will help an archaeologist or forensic scientist to establish the gender of human remains.

Peak performance in activities needing physical strength and flexibility rely on a system of soft tissues supported by a tough and flexible **skeleton**. **Muscles** provide movement at joints and maintain posture.

As well as supporting your body, your skeleton:
- stores minerals such as calcium and phosphorus
- makes red blood cells, platelets, and some white blood cells in bone marrow
- forms a system of levers with muscles attached, which allows the body to move
- protects internal organs, for example, the pelvis protects the reproductive organs.

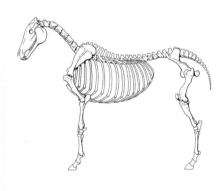

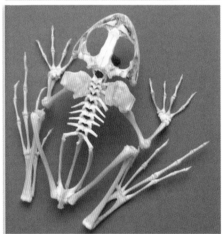

Vertebrates have bones inside them – match the creature to the skeleton.

Living bone

The skeleton is not just dry bone. Its tissues, such as **bone** and **cartilage**, are made of living cells. Blood brings nutrients and oxygen to the cells.

Bone is continually broken down and rebuilt, which allows a child's bones to grow in size. Even an adult's skeleton is continually changing. Weight-bearing exercises such as jogging stimulate bone growth, increasing its density. Inactivity makes bone less dense and weaker.

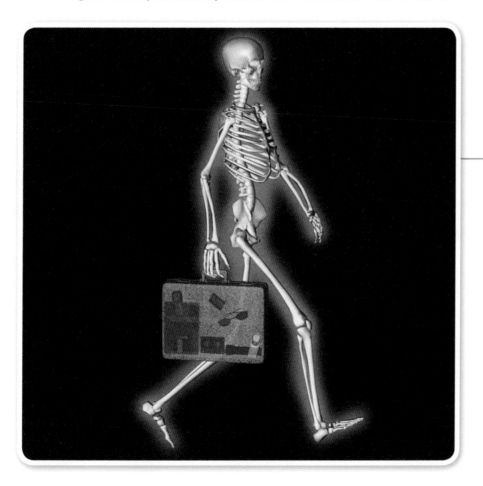

Key words

- ✓ **skeleton**
- ✓ **muscle**
- ✓ **bone**
- ✓ **cartilage**

Questions

1 List four functions of the skeleton.

2 Describe how exercise changes bones.

The human skeleton has over 200 bones. Most will move, but some are fixed in position, for example, those in the skull. Skull bones are flexible during early development, but fuse together soon after birth.

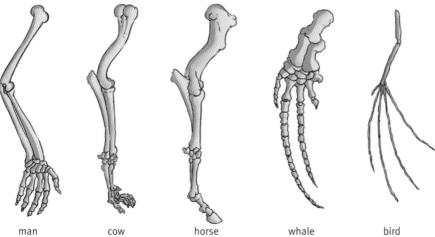

| man | cow | horse | whale | bird |

It is interesting to think that all vertebrate skeletons are developed from the same original plan. Can you spot the same bones in each creature? A horse walks round on single 'fingers' with big 'finger nails' as hooves.

Holding the bones together

Any sport or physical activity involves movement where two or more bones meet at a joint. Ball-and-socket joints, at your hip and shoulder, are the most versatile. These joints move in every direction, like a computer joystick. Hinge joints, such as the knee and elbow, move in just two directions – backwards and forwards.

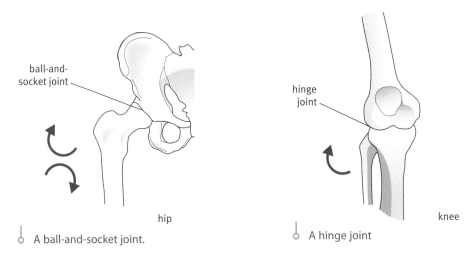

ball-and-socket joint

hip

⊙ A ball-and-socket joint.

hinge joint

knee

⊙ A hinge joint

Tough elastic bands, called **ligaments**, hold the bones in place and limit how far the bones can move. Smooth cartilage stops bones from grinding against each other as they move. It forms a rubbery shock-absorbing coat over the end of each bone. This stops the bones from damaging each other. Cartilage is smooth, but friction could still wear it down. To reduce this as much as possible, the joint is lubricated with oily **synovial fluid**.

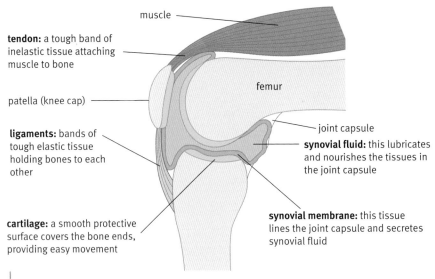

muscle

tendon: a tough band of inelastic tissue attaching muscle to bone

patella (knee cap)

ligaments: bands of tough elastic tissue holding bones to each other

cartilage: a smooth protective surface covers the bone ends, providing easy movement

femur

joint capsule

synovial fluid: this lubricates and nourishes the tissues in the joint capsule

synovial membrane: this tissue lines the joint capsule and secretes synovial fluid

⊙ The knee joint. Like most joints in the body, this is a synovial joint.

How muscles move bones

Inelastic **tendons** transmit the forces from muscles to the bones. Muscles can only pull a bone for movement at a **joint**. After contracting, the muscle is stretched again only when the bone is pulled back by another muscle. So at least two muscles must act at every joint:

- One muscle contracts to bend the joint.
- The other muscle contracts to straighten it.

Muscles that work opposite each other are called an **antagonistic pair**.

There are over 600 muscles attached to the human skeleton. They make up almost half of your total body weight.

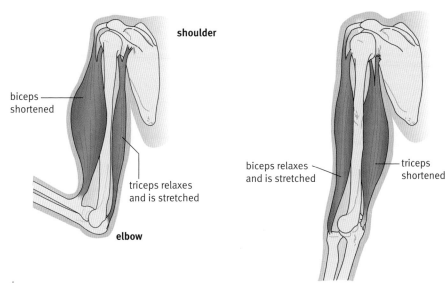

shoulder

biceps shortened

triceps relaxes and is stretched

elbow

biceps relaxes and is stretched

triceps shortened

The biceps and triceps muscles contract to move the elbow joint.

Questions

1 Describe the difference between a tendon and a ligament.

2 Explain why this difference is important.

3 Name the parts of a synovial joint and explain how they are suited to their function.

4 Name the muscle that:
   a bends the arm
   b straightens the arm.

5 Explain what is meant by an antagonistic pair of muscles.

6 Professional dancers and gymnasts often develop osteoarthritis in their knee joints in later life. The surface of their cartilage becomes rough and wears away. Suggest what symptoms this will cause.

Find out about

- ✓ **what you need to consider before starting an exercise programme**
- ✓ **measuring how your body changes when you exercise**

Circuit training is a good way to build up your strength and fitness level at a pace to suit you. The instructor can vary the difficulty of the exercises or the number of times that they have to be performed.

Scuba divers need to have a particularly thorough medical examination before they can train to dive with their club. They will need to have a chest X-ray and a doctor's signature in their diving log book.

Starting out

When you join a new sports club your instructor should find out some key facts about you. You will be asked to fill out a questionnaire about your **medical** and **lifestyle history**. This will help your instructor to plan an exercise regime that suits you.

If you start a new activity you should tell your instructor about your medical history and lifestyle. What adjustments might you need to make to these people's exercise regimes?

Medical history

Before starting a new exercise regime, your instructor might enquire if you, or anyone in your family, suffers from certain medical conditions. Your regime will be adjusted to take into account any circulatory and respiratory problems. Your instructor needs to know about any treatment you have had for injuries – especially to your joints or back. Let your instructor know of any **medication** you take. You might need to have medication, such as an inhaler, close at hand.

Lifestyle history

The coach might also ask questions about your lifestyle. If you have a healthy and active lifestyle you will progress quicker than someone who is new to the activity or who has been resting while recovering from an injury. The duration and difficulty of exercise will be adapted to suit your needs.

Getting fitter

Baseline data

When you begin a new activity you might take some initial measurements. This is your **baseline data**. The next time that you monitor yourself you will be able to see how much you have improved by.

Heart rate: As you exercise your heart rate increases. Your heart beats faster to deliver more food and nutrients to your muscles. It is recommended that you train at about 60% of your maximum heart rate.

Blood pressure: When you do strenuous exercise your heart beats more forcefully and your **blood pressure** increases. 120/80 mmHg is a typical value for a healthy person.

Recovery period: The fitter you get the faster your **recovery period** after physical activities.

Proportion of body fat: Too much body fat puts a strain on your heart and your arteries may become dangerously narrowed.

Body mass index (BMI): This measures your body mass based on your height. Use this formula to work out your **body mass index**.

$$BMI = \frac{\text{body mass (kg)}}{[\text{height (m)}]^2}$$

BMI tables indicate if your body mass is healthy for your size. For adults, a BMI between 18.5 and 24.9 is considered healthy.

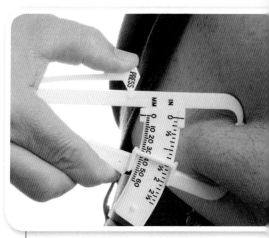
Callipers measure how much fat lies under your skin.

Key words

- ✓ medical history
- ✓ lifestyle history
- ✓ medication
- ✓ blood pressure
- ✓ recovery period
- ✓ body mass index (BMI)

Questions

1 If you worked in a fitness club, what health questions might you ask new clients when they joined?

2 Outline three different fitness tests that could be carried out. For each one, describe how the measurements would change as the person got fitter.

3 The table shows the heart rates of three people taken after five minutes of exercise. The measurements were repeated two months later.

Heart rate in beats per minute (bpm) after five minutes of exercise

Person	Baseline measurements (bpm)	Measurements two months later (bpm)
A	148	122
B	161	158
C	132	152

i Who has been resting following a sports injury?

ii Who has not done much exercise?

Measurement	What happens after exercise as you get fitter
Heart rate	Your pulse will not beat so quickly.
Blood pressure	Your blood pressure will reduce.
Recovery period	The time for you to recover will get shorter.
Proportion of body fat	The proportion of body fat will get less.
Body mass index	Your BMI value will fall as your mass decreases.

How will your results change as you get fitter?

Find out about

✔ **the best way to gather accurate, reliable fitness measurements**

Try to take your fitness measurements as accurately as possible. Damaged or defective equipment will not provide the measurements you need.

Key words

✔ **accurate**
✔ **calibrated**
✔ **repeatable**

Questions

1 Explain the difference between accuracy and repeatability in your own words.

2 Describe three ways that you could mistakenly calculate that you are fitter than you actually are.

Accurate data

An **accurate** instrument or procedure gives a 'true' reading. Some of these measurements need equipment to take them, but the equipment could be faulty. Different instruments can give different readings – especially if they have been subjected to rough use. Double check your measurements with different equipment.

Equipment used by medical and sports professionals will be **calibrated** regularly. This means that the measurement from the equipment under test is compared to the measurement of equipment that is known to be of the correct standard.

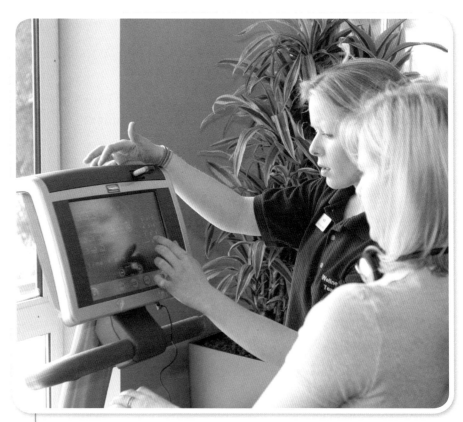

Measuring equipment should be checked and calibrated regularly to make sure that it is accurate.

Repeatable data

Scientists need data that can be trusted. If your data is **repeatable** you get similar results with each re-run of an experiment. Repeatable results suggest that you can have confidence in your techniques and methods. Take time and care when taking measurements; try to get a 'feel' for what the values should be, so that you can spot errors quickly.

Heart rate

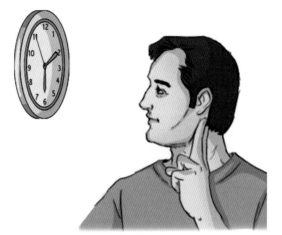

It is hard to measure your pulse rate when you have been exercising very hard. Work with a partner, but make sure they can find your pulse. Feel for the pulse with your fingers only – your thumb has its own pulse, which might confuse you.

Blood pressure

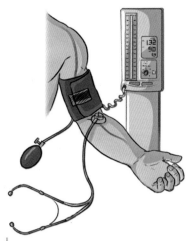

Make sure the measuring equipment is not damaged or has low battery power. Always measure your blood pressure while sitting down with your arm at chest height. Raising your arm will lower the pressure (which is why you should elevate a wound if it is bleeding). Your blood pressure can increase if you are feeling tense or angry. Recognise that the values will be higher if you are stressed.

Recovery period

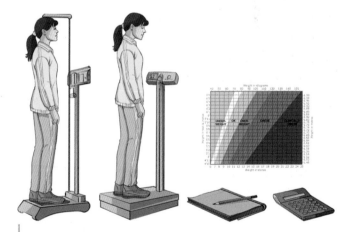

Double check your calculations and use the average of several measurements. Don't be put off if it seems that your fitness measurements are only improving slowly.

Body fat

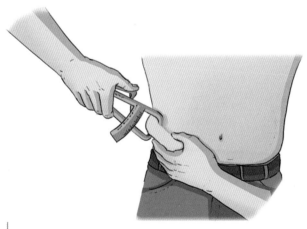

Take care when reading the scales on any measuring devices, particularly if they are worn. Look to see that you are reading the right units (some old-fashioned equipment may have imperial units). Take the reading at the right place on the scale.

Body mass index (BMI)

Your mass will often vary at different times of the day depending on whether you have just eaten and how dehydrated you are. Always measure your mass at the same time of day to compare like with like. When you refer to graphs or tables of data, make sure you are using a reputable source. Remember, different fitness graphs are prepared for men, women, and children.

Find out about

✔ injuries caused by excessive exercise

Key words

✔ sprain
✔ dislocation
✔ torn ligament
✔ torn tendon
✔ RICE

Joint injuries

If you follow a sports team, you will know how often players may get injured. It is an occupational hazard. Joints are tough and well-designed, but there is a limit to the force they can withstand. Common injuries include **sprains**, **dislocations**, **torn ligaments**, and **torn tendons**.

Football is particularly hazardous. There are lots of stops, starts, and changes of direction, and perhaps some bad tackles. It is not just professional footballers who suffer – 40% of knee injuries happen to under-15 footballers.

Sprains

The most common sporting injury is a sprain. This usually happens when you overstretch a ligament by twisting your ankle or knee. Often people will say that they have 'torn a muscle', when they have actually sprained a ligament. There are several symptoms:

• redness and swelling
• surface bruising
• difficulty walking
• dull, throbbing ache or sharp, cramping pain.

The usual treatment for sprains is **RICE** – rest, ice, compression, elevation.

REST means immobilising the injured part (e.g. keeping the weight off a torn muscle).

ICE acts as an anaesthetic, reduces swelling, and slows the flow of blood to the injured area. To avoid damaging the tissue, the ice is applied indirectly (e.g. in a tea towel or plastic bag) for up to 20 minutes at a time with 30 minutes between applications.

COMPRESSION usually involves wrapping a bandage round the injured part to reduce swelling. The bandage should be snug but not too tight.

AEROBIC EXERCISING of the injured part is not restarted until it has regained at least 75% of the previous level of strength, and then only moderately. This exercise helps build muscle and return the athlete to peak fitness.

SIMPLE STRETCHING ROUTINES help to regain mobility, but only when swelling stops.

ELEVATION means raising the injured limb. This reduces swelling by helping to keep excess fluid away from the damaged area.

Recovery from a sports injury often involves RICE followed by stretching and strengthening exercises. RICE stands for rest, ice, compression, and elevation.

After 72 hours of RICE treatment, heat and gentle massage can be used to loosen the surrounding muscles. If the injury keeps occurring, physiotherapy can be used to strengthen the surrounding muscles.

Torn ligaments and tendons

If a joint is twisted or overstretched then the ligaments and tendons attached to it may tear.

A ligament may tear with a popping sound, leaving the joint painfully bruised and very hard to bend. You might be able to see a dent where the ligament is torn. Tendons can stretch, become inflamed, and even snap like a worn-out elastic band. This might happen in sports involving a lot of jumping – like basketball.

If you tear a ligament, use the RICE treatment and see a physiotherapist. Don't do vigorous exercise for two to three months. In severe cases the joint might have to be immobilised with a brace or repaired by surgery. It is best to avoid injury in the first place by building up gradually to exercise and resisting over-exertion.

Dislocations

Gymnasts also suffer from joint injuries, often at the knee joint. Cartilage in the knee is an excellent shock absorber, but floor routines put a lot of force on the joints. If a gymnast lands off balance, their kneecap can become dislocated. This happens when the bone slips out of the joint. In contact sports such as rugby, dislocations of the shoulder are extremely common. Dislocations are very painful.

The Royal Navy Field Gun race claims many injuries.

Gymnasts land with great force. It can be equivalent to carrying ten times their body weight.

Questions

1 Describe:
   a the symptoms and
   b RICE treatment for a sprain.

2 Describe two other types of common joint injury.

Joint injuries are common in football players.

Mike's experience of physiotherapy

'I went running with my son Joe. He set a quick pace and the next day my knee really hurt. The pain didn't go away. When I went to the doctor, she prescribed me some strong anti-inflammatory tablets and referred me to the hospital physiotherapist. By the time I saw the physiotherapist my knee felt much better. After giving me a thorough examination, the physiotherapist explained everything carefully and gave me a list of exercises to do. These seemed a bit dull and repetitive, so I didn't bother much. Two weeks later I was back at the doctors with my knee as bad as ever – this time I'll follow the physiotherapist's advice!'

What the physiotherapist says

'That's quite a typical story for our middle-aged patients,' says Vicky Singleton. At 25 she is a senior physiotherapist – a professional who assesses how bad an injury is and treats it through physical exercises and manipulation. It takes a lot of training to qualify as a physiotherapist. Vicky had to get good A-level results (particularly biology) and then study for three years to get her degree. As part of her training she spent 1000 hours working with other physiotherapists and practising her skills on patients.

Vicky has seen lots of patients like Mike, with problems in their joints or muscles.

How can physiotherapy help?

Physiotherapy can be used immediately to treat sporting injuries, starting with RICE. From then on, physiotherapy is used to encourage healing and return a joint or muscle to full strength and movement.

Find out about

- ✔ **the role of a physiotherapist in the treatment of skeletal–muscular injury**
- ✔ **the need to comply with a physiotherapy treatment regime**

It's easy to damage your joints when you exercise.

'We use a series of gentle exercises to begin with,' Vicky explains. 'These make sure that the tissues heal well and don't shrink or tighten too much. Lots of rest is important too. Then we give more exercises, which strengthen the new tissues and make sure the joint moves as well as ever. It can take six to twelve weeks for the tissues to heal and then probably another six weeks before the joint is fully back to strength. Really bad injuries can take up to two years of physiotherapy to put them right!'

Speeding up recovery

Mike's injury wasn't very serious. The damage that footballers and other professional athletes do to their bodies can be much worse. They may try to get active again quickly by using injections to reduce the swelling, speed up healing, and kill the pain. They will also have physiotherapy several times a day. This often works well – but the problem with rushing a recovery is that the joint tissues may not be fully healed before they are used, so even more damage can result.

Keeping up the treatment

Physiotherapy involves repeating exercises many times to build up strength and movement.

Some sporting injuries are very severe and can take months to heal.

A good physiotherapist makes sure the patient understands what they need to do and why they need to do it. If patients don't comply with their treatment, like Mike, the tissues don't heal properly, and the joint may become stiff and painful permanently.

Physiotherapy doesn't just help skeletal and muscular injuries. It can be used to help children and adults with disabilities gain strength and control over their bodies.

Vicky is specially trained to do physiotherapy with disabled children on horseback. This helps their muscles and joints become stronger and more flexible, improving their balance and quality of life.

Questions

1 What are the advantages of physiotherapy when you have injured a joint?

2 Why is it important to comply with the treatment your physiotherapist gives you?

Find out about

- ✔ the circulation system
- ✔ what blood does and what it contains

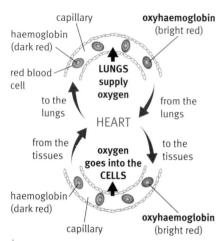

In the lungs oxygen is at high concentration and binds to **haemoglobin**. At low concentration in body tissues the oxygen is released. It diffuses into body cells, which use the oxygen for respiration.

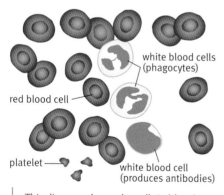

This diagram shows the cells in blood, but not in the correct proportions. Each 1 mm³ of blood contains approximately 5 million red blood cells, 250 000 platelets, and 7000 white blood cells.

Systems for moving molecules

Peak performance of your body relies on an efficient circulation system. It would take too long for oxygen and other molecules to diffuse from outside our bodies to the millions of cells inside. Active muscles need a good supply of glucose and oxygen. Carbon dioxide and other waste products need to be removed.

The circulatory system

When you are at rest, your heart beats about 60 to 80 times per minute, forcing blood through a network of blood vessels. Arteries carry the blood to your organs. In fine capillaries, useful molecules pass from the blood into the cells. Waste products are taken away. Veins return the blood to the heart.

Double circulatory system

Because the capillaries are so tiny, it was only a few hundred years ago that doctors first saw them. They proved that blood went round the body and back to the heart in a loop. In fact our blood is pumped around two separate loops: around the body and around the lungs. This is called double circulation.

A double pump

As we have a double circulatory system we need a heart that is two pumps in one. Blood needs to be at a high enough pressure to be carried all round our bodies. Lower pressure in the blood sent to the lungs stops fluid from being forced out into our airways.

Blood

Plasma

You have between five and seven litres of blood circulating around your body. A sample of blood looks completely red. A closer look shows that it consists of cells floating in pale yellow liquid called **plasma**. Plasma is mainly water. It transports a wide range of materials including nutrients such as glucose, antibodies, hormones, and waste such as carbon dioxide and urea. Plasma gives blood its bulk and also helps to distribute heat around the body.

There are three types of cells floating in plasma:

- **red blood cells** – to transport oxygen
- **white blood cells** – to fight infection
- **platelets** – which play an important part in blood clotting at an injury site.

Red blood cells

Red cells are the most obvious blood component because of their number and colour. The cells are adapted to their function of transport. They are packed with the protein haemoglobin. Haemoglobin binds oxygen as blood passes through lungs. The oxygen is released from haemoglobin as blood circulates through the tissues of the body.

Red blood cells have no nucleus, which allows more space for haemoglobin. If the haemoglobin circulated freely in the plasma instead, then the blood would be too thick to flow properly.

The biconcave shape gives the cells a large surface area, making diffusion of gases very efficient. This shape also gives cells flexibility to squeeze through tiny capillaries.

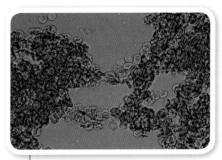

Clotted red blood cells.

White blood cells

White blood cells protect the body from infection by disease-causing microorganisms. They produce antibodies, and engulf and digest microorganisms by **phagocytosis**. You can read more about white blood cells and immunity in module B2.

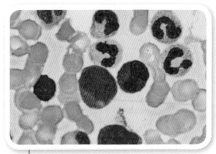

White blood cells.

Platelets

Platelets are fragments of cells that are made from the cytoplasm of large cells. When a blood vessel is damaged, for example, when you are cut, platelets stick to the cut edge. They send out chemicals that trigger a series of reactions that form a clot at the cut site. Clotting helps stop too much blood being lost from the body.

Key words

- ✓ **plasma**
- ✓ **red blood cells**
- ✓ **white blood cells**
- ✓ **platelets**
- ✓ **phagocytosis**

Questions

1 Simple single-celled pond organisms rely on diffusion for exchanging food, oxygen, and waste with the surrounding water. Explain why you need a circulation system whereas simple pond creatures do not.

2 Copy and complete the following table for the four main components of blood.

Blood component	Appearance	Function

Find out about

✓ **the structure of the human heart**

A double circulation

Each side of the heart has two chambers – an **atrium** and a **ventricle**.

Blood from the body enters the right atrium of the heart. It is pumped out of the right ventricle towards the lungs to pick up oxygen. The blood becomes **oxygenated**. It returns to the left atrium and passes into the left ventricle. Here it gets another, harder pump, which carries it around the rest of the body. The left ventricle has a thicker wall of muscle than the right, because it has to pump blood to the whole of the body. The right ventricle only pumps blood as far as the lungs. As the blood passes around the body it gradually gives up its oxygen to the cells. It becomes **deoxygenated**. The blood then returns to the right atrium again. So blood passes through the heart twice on every circuit of the body. This is called a **double circulation**.

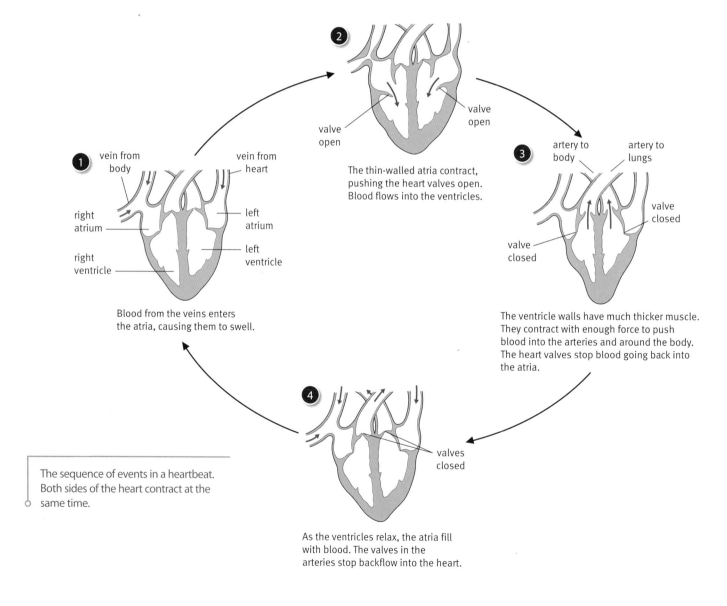

1 vein from body · vein from heart · right atrium · right ventricle · left atrium · left ventricle

Blood from the veins enters the atria, causing them to swell.

2 valve open · valve open

The thin-walled atria contract, pushing the heart valves open. Blood flows into the ventricles.

3 artery to body · artery to lungs · valve closed · valve closed

The ventricle walls have much thicker muscle. They contract with enough force to push blood into the arteries and around the body. The heart valves stop blood going back into the atria.

4 valves closed

As the ventricles relax, the atria fill with blood. The valves in the arteries stop backflow into the heart.

The sequence of events in a heartbeat. Both sides of the heart contract at the same time.

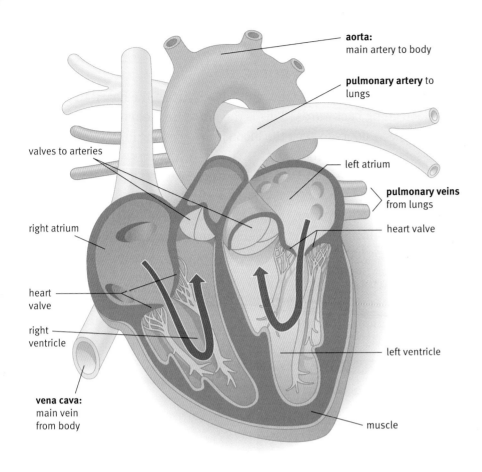

aorta:
main artery to body

pulmonary artery to
lungs

left atrium

valves to arteries

pulmonary veins
from lungs

right atrium

heart valve

heart
valve

right
ventricle

left ventricle

vena cava:
main vein
from body

muscle

Diagram of a human heart. 'Left' and 'right'
refer to the way your heart lies in your body.
The artery leaving the right-hand side of the
heart has been coloured blue to show that it
is carrying blood that is short of oxygen.

Key words

- ✓ **atrium**
- ✓ **ventricle**
- ✓ **oxygenated**
- ✓ **deoxygenated**
- ✓ **double circulation**
- ✓ **aorta**
- ✓ **vena cava**
- ✓ **pulmonary vein**
- ✓ **pulmonary artery**

Questions

1 Explain what is meant by a
double circulatory system.

2 Explain the difference in
wall thickness between:
   a atria and ventricles
   b the right and left
     ventricles.

capillaries of
head and neck

capillaries
of lungs

vena cava
(main vein from
body)

pulmonary
artery

pulmonary veins
from left lung

vena cava
(main vein
from body)

aorta (artery to body)

hepatic artery to liver

hepatic vein

artery to stomach

capillaries of
liver

capillaries of
intestines

renal vein

renal artery to kidney

capillaries of
kidney

capillaries of
lower body

Blood is carried away from the heart by arteries,
and towards the heart in veins. Trace with your
finger the different routes that a red blood cell
might make around the body. You can read more
about the structure and function of blood vessels
in module B2.

Find out about

- ✓ **the function of valves in the heart and veins**
- ✓ **capillary beds, where nutrients and waste are exchanged**
- ✓ **why tissue fluid is important**

Valves – a one-way system

When the ventricles contract they push blood out of the heart. But what stops blood going backwards? This is the job of the **heart valves**. They act like one-way doors to keep the blood flowing in one direction. There are two sets of valves in the heart:

- between each atrium and ventricle — these valves stop blood flowing backwards from the ventricles into the atria
- between the ventricle and the arteries leaving the heart — these valves stop blood flowing backwards from the arteries into the ventricles.

Valves are also found in veins. Blood pressure is lower in veins than in arteries. Valves stop blood from flowing backwards in the veins in between each pump from the heart.

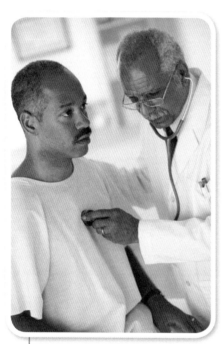

The comforting 'lub-dub' sound of a heartbeat is the sound of heart valves snapping shut. The 'lub' sound is caused by the valves between the atria and ventricles shutting. The 'dub' sound is made as the valves between the ventricles and the arteries close.

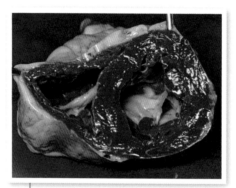

A valve between a ventricle and an artery. Strong tendons hold the valve flaps in place, preventing blood from flowing backwards.

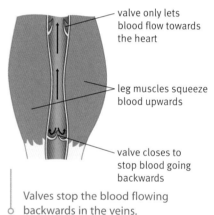

valve only lets blood flow towards the heart

leg muscles squeeze blood upwards

valve closes to stop blood going backwards

Valves stop the blood flowing backwards in the veins.

Capillary beds

The lungs and other body organs contain a bed or network of **capillaries**. As blood squeezes through these capillaries its pressure drops. It has less force moving it along. If blood went straight from the lungs to the rest of the body, it would move too slowly to provide enough oxygen for the body's cells. The double circulation gives the blood a pump to get through each bed of capillaries – the right ventricle pumps it through the lungs, and the left ventricle pumps it through the rest of the body.

Key words

- ✓ **heart valves**
- ✓ **capillary**
- ✓ **tissue fluid**
- ✓ **capillary bed**

Diffusion in the capillary beds

On average you have six litres of blood in your body. It is all pumped through the heart three times each minute. But the blood spends most of that time in capillary networks. Capillaries are where chemicals in the body's cells and in the blood are exchanged. The structure of capillaries makes them ideally suited for this function.

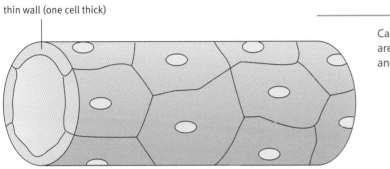

thin wall (one cell thick)

Capillary walls are very thin and porous.

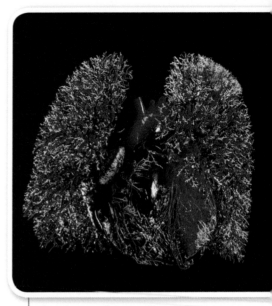

A resin cast of the capillary network of an adult human's lungs.

Tissue fluid

When blood enters a capillary bed from an artery it is at high pressure. Blood plasma is squeezed out of the capillary. It forms a liquid called **tissue fluid**, which bathes all of your cells.

Tissue fluid contains all the dissolved raw materials being carried by blood plasma, including glucose and oxygen. These chemicals diffuse from the tissue fluid into cells. Waste products from cells, including urea and carbon dioxide, diffuse out into the tissue fluid.

As blood passes through the capillary bed its pressure drops. Plasma stops being squeezed out, and tissue fluid with waste products from cells moves back into the capillaries.

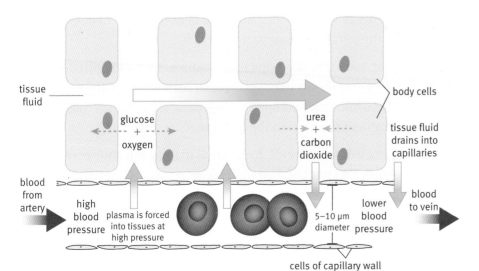

Tissue fluid helps nutrients and waste products to be exchanged by diffusion in the **capillary beds**.

Questions

1 Draw a flow chart to show the route blood takes on its journey around the body. Include the key words on pages 211–214 in your description. Start with: Blood leaves the left ventricle.

2 Describe the job of valves in the heart and veins.

3 What is tissue fluid made from?

4 Explain why tissue fluid leaks out at the start of capillaries, and moves back in towards the end of the capillary bed.

5 Explain how tissue fluid helps the exchange of chemicals between the capillaries and body tissues.

6 Name four chemicals that are exchanged between body cells and tissue fluid.

Find out about

- ✔ **why you must balance heat gain and heat loss**
- ✔ **how your body detects and responds to temperature changes**
- ✔ **how sweating and being 'flushed' help to cool you down**
- ✔ **how shivering and turning pale may stop you getting too cold**

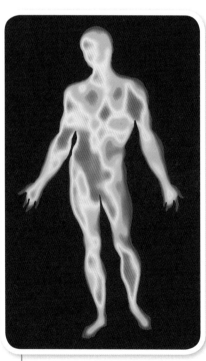

Our body radiates heat. On this thermal image, the hottest parts of the body are red and the coolest blue.

Key words
- ✔ **respiration**
- ✔ **extremity**
- ✔ **core**

For peak performance, your body needs to keep a constant internal environment. Enzymes are catalysts, speeding up chemical reactions in cells without taking part. Enzyme reactions need to work at their optimum temperature for maximum efficiency.

Gaining and losing heat

If your environment is hotter than you are, energy will be transferred to your body. Your body also gains heat from **respiration**. During respiration, glucose is broken down to release energy for cells. This energy is used by muscles for movement. Some of the energy warms up the body.

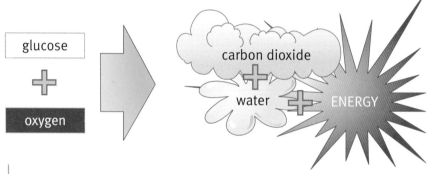

Respiration releases energy from the breakdown of glucose.

If your environment is cooler than you are, energy will be transferred away from your body. The bigger the temperature difference, the greater the rate of cooling.

Getting the balance right

For your temperature to remain constant, energy gain must be balanced by energy loss. In other words, if heat gained = heat lost, your body temperature stays the same.

Not all of your body is at the same temperature. Your **extremities** (hands and feet) are cooler than your **core** (deeper parts). Extremities have a larger surface area compared with their size. So they lose energy to the environment faster than the main parts of your body.

The temperature of your hands, feet, and skin falls, but your core temperature hardly changes. It should be between 36 °C and 37.5 °C for your body to work properly. 'Normal' body temperature varies from one person to the next.

The amount of energy released in respiration and other reactions is greatest in your muscles and liver. The circulation of your blood transfers this energy to other parts.

Investigating temperature control

A physiologist called Sir Charles Blagden was Secretary of the Royal Society towards the end of the 18th century. Like many physiologists, he experimented on himself. He went into a very hot room to see how his body would react. The account below is adapted from Harry Houdini's book, *The Miracle Mongers: An Exposé*.

Another account describes Blagden taking a dog into the room with him, along with steak and eggs. The dog was unharmed but had to stay in its basket so it did not burn its feet on the hot floor.

Blagden's experiment

Sir Charles Blagden went into a room where the temperature was 1 degree or 2 degrees above 127 °C, and remained eight minutes in this situation, frequently walking about to all the different parts of the room, but standing still most of the time in the coolest spot, where the temperature was above 116 °C. The air, though very hot, gave no pain, and Sir Charles and all the other gentlemen were of the opinion that they could support a much greater heat.

During seven minutes Sir Charles´s breathing continued perfectly good, but after that time he felt an oppression in his lungs, with a sense of anxiety, which induced him to leave the room. His pulse was then 144 [beats per minute], double its ordinary rate. In order to

Blagden's dog takes part in the experiment.

prove that the thermometer was not faulty, they placed some eggs and a beef-steak upon a tin frame near the thermometer, but more distant from the furnace than from the wall of the room. In twenty minutes the eggs were roasted quite hard, and in forty-seven minutes the steak was not only cooked, but almost dry.

Questions

1 What two things must be balanced to keep your body temperature steady?

2 What part of your body is warmest?

3 At what temperature does this warmest part need to be maintained?

4 Why is your blood important in keeping extremities warm?

5 What effects did the very high temperatures have on Sir Charles Blagden?

6 What happened to the proteins in the steak and eggs as they cooked?

7 Why did the same thing not happen to Sir Charles's proteins?

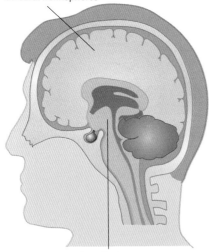

cerebral hemispheres

hypothalamus

The **hypothalamus** is the **processing centre** in the brain for sleep, water balance, body temperature, appetite, and other basic functions. The cerebral hemispheres are where you make conscious decisions to warm or cool yourself.

The fans in this Sikh temple provide a welcome breeze to help to keep the people cool.

Key words
- ✓ hypothalamus
- ✓ processing centre
- ✓ shivering
- ✓ sweat glands
- ✓ evaporate

Changes in your body temperature can have serious effects on health. Your brain is particularly sensitive to temperature changes. Temperature receptors in your skin can detect a change in air temperature of as little as 0.5 °C. There are temperature receptors in the area of the brain called the **hypothalamus** that detect blood temperature. The hypothalamus is also the **processing centre** for temperature control. When the temperature in your brain is above or below 37 °C, the hypothalamus triggers effectors to bring your body temperature back to normal.

Warming up

Shivering is one way your body keeps warm. When you shiver, muscle cells contract quickly. They respire faster to release the energy for this movement.

Shivering is an automatic response. You may also take a conscious decision to do something that will warm you up, for example, drinking a warm drink, putting on more clothes, or going inside.

Cooling down

When you are too hot, nerve impulses from the brain stimulate your **sweat glands**. Sweat passes out of small pores onto the skin surface. Water molecules in sweat gain energy from your skin. Soon they move quickly enough to **evaporate**. This cools you down. Even when it is cool and you are not very active, you can lose nearly a litre of water a day in sweat. When you are hot and active, you can lose up to three litres of water an hour. If you don't replace this water you could become dehydrated. Dehydration then reduces sweating, raising the body temperature further. This is what happens when you get 'heat stroke.'

How does sweat cool you down?

Sweating only works to cool you down if the sweat can evaporate quickly. In a hot, humid climate sweat drips off you and you feel uncomfortably hot. Air currents increase the rate of evaporation and so increase the cooling effect.

Is your body temperature the same all day?

All these responses keep your body temperature within a narrow range. Average core body temperature is about 36.9 °C, but this varies from person to person and it varies throughout the day. For example, when you are sleeping, you move around less and respire more slowly. So your body temperature drops to its lowest point at night.

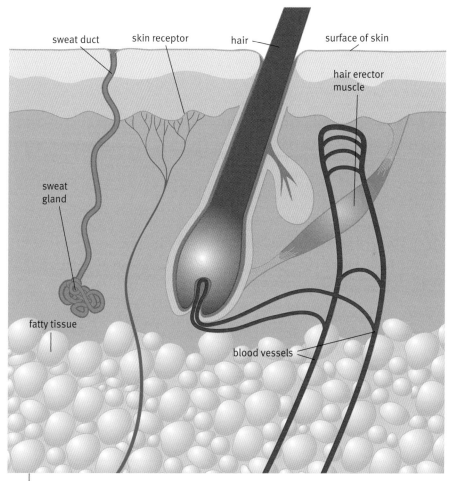

The structure of skin.

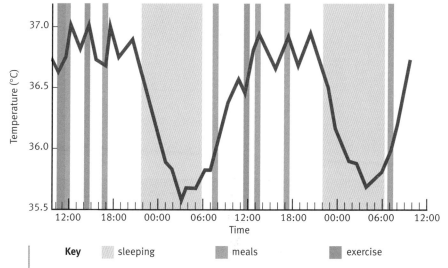

Key ▨ sleeping ▨ meals ▨ exercise

The daily cycle of variation in body temperature. You can see that eating, sleeping, and activity affect body temperature. These fluctuations happen even when you are at rest – they are controlled by our 'biological clocks'.

In hot, humid conditions, sweat cannot evaporate easily. Your clothes may become soaked in sweat.

Questions

1 Where in your body would you find:
   a temperature receptors?
   b the temperature processing centre?

2 Name two effectors for controlling body temperature.

3 Explain how shivering warms you up.

4 Explain how sweating cools you down.

Too hot or too cold?

The table below lists ways of controlling body temperature. Some of the effectors warm you up and some cool you down – they are **antagonistic**.

Too cold?	Too hot?
Muscles shiver Energy warms the body tissues as muscles contract when you shiver.	**Muscles don't shiver**
No sweat made	**Sweat produced** Energy is lost from the skin, when the water molecules in sweat evaporate.
Hair raising Erector muscles make your hair stand on end, trapping a layer of warm air.	**Hair lies flat**
Skin goes paler Blood is diverted away from the skin surface to the core to retain warmth (**vasoconstriction**).	**Skin becomes flushed** Blood vessels near the skin surface widen, helping heat to be lost to the surroundings (**vasodilation**).
Warming behaviours You could do exercises to warm up. Putting on more clothes will reduce heat loss. Moving to a warm place or having a hot drink will warm you up.	**Cooling behaviours** Taking a break from exercise or moving to the shade will let your body temperature recover. Removing clothes increases the rate of heat loss. If you wet your skin or fan yourself, evaporation will provide a cooling effect.

Question

1 For each picture on this page, say how the action cools the body.

Vasoconstriction and vasodilation

The boy on the right is flushed as a result of vasodilation. 'Vaso' is from the Latin for vessel, so vasodilation means widening of the blood vessels. More blood flows into the capillaries in the skin, so there is more energy transfer to the environment. The opposite is vasoconstriction. Less blood reaches the capillaries in the skin, so energy loss is reduced.

This boy has a fever. He is sweating and looking flushed.

Vasodilation
The blood vessels near the surface of the skin are filled with blood. Energy from the warm blood is transferred down the temperature gradient to the environment.

Vasoconstriction
The muscles in the walls of blood vessels near the surface of the skin contract. Less blood flows near the surface of the skin, so less energy is lost to the environment.

surface of skin

blood vessels near surface of skin

Key ⟶ flow of blood in blood vessels ⟿ energy loss from skin surface

Vasodilation and vasoconstriction are a good example of the effects of control by antagonistic effectors.

Key words
- ✓ antagonistic
- ✓ vasodilation
- ✓ vasoconstriction

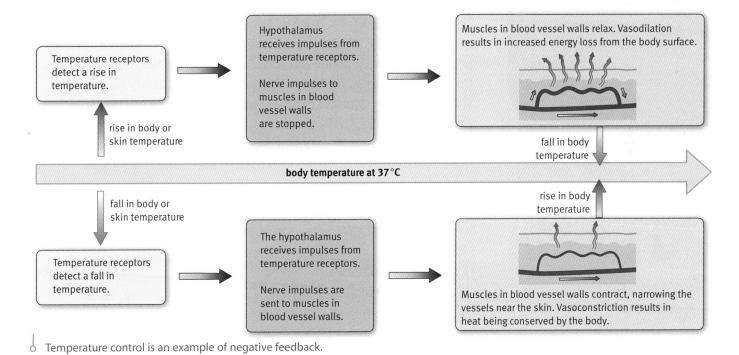

Temperature receptors detect a rise in temperature.

rise in body or skin temperature

Hypothalamus receives impulses from temperature receptors.

Nerve impulses to muscles in blood vessel walls are stopped.

Muscles in blood vessel walls relax. Vasodilation results in increased energy loss from the body surface.

fall in body temperature

body temperature at 37 °C

fall in body or skin temperature

rise in body temperature

Temperature receptors detect a fall in temperature.

The hypothalamus receives impulses from temperature receptors.

Nerve impulses are sent to muscles in blood vessel walls.

Muscles in blood vessel walls contract, narrowing the vessels near the skin. Vasoconstriction results in heat being conserved by the body.

Temperature control is an example of negative feedback.

Questions

2 What are the effectors that cause vasodilation and vasoconstriction?

3 Explain why vasodilation and vasoconstriction are said to be controlled by antagonistic effectors.

4 Why does vasodilation not cool you when the temperature of the environment is higher than your body temperature?

Some diabetics wear a wrist band indicating that they have diabetes. Why do you think they do this?

Sugar for energy

Your cells need sugar to give them the energy to function.

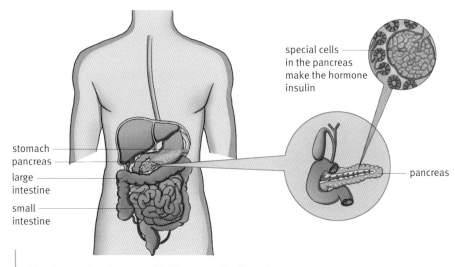

special cells in the pancreas make the hormone insulin

stomach
pancreas
large intestine
small intestine

pancreas

Blood sugar level is controlled by special cells in the **pancreas**.

Steady sugar release

Many processed foods have extra sugar added to them to improve their taste. A lot of sugar in one go causes a surge of the **hormone insulin** that removes it from the blood. The energy boost quickly turns into a sugar low, causing lack of energy, concentration, and mood swings. Foods that are high in fibre and complex carbohydrates like pasta, rice, and bread keep your blood sugar level balanced. These foods are digested slowly. Their sugars are released gradually into your blood stream, which is better for you.

People with diabetes cannot control their blood sugar level. There are two main types of **diabetes: type 1 and type 2**.

Type 1 diabetes

Some people develop type 1 diabetes when they are young. Their pancreas suddenly stops making enough insulin. Blood sugar levels could rise dangerously high, particularly after a sugary meal. Symptoms include thirst and the production of large volumes of urine containing sugar. Too high a level of blood sugar makes you drowsy, too low, and you may go into a coma.

People with type 1 diabetes need several daily injections of insulin to control their blood sugar level. They have to be careful about what they eat, to match their sugar intake to their lifestyle. They may have to prick their finger each day and measure their blood sugar level using a special test strip.

Type 2 diabetes

People with a poor diet, inactive lifestyle, or who are **obese** may develop type 2 (or 'late onset') diabetes in middle age. The number of young people with type 2 diabetes is rising. In type 2 diabetes the body gradually stops making enough insulin for your needs or the cells can't use the insulin properly. Symptoms include thirst, frequent urination, tiredness, and weight loss. Over time 'hardening of the arteries' takes place, which can lead to heart attacks, kidney damage, or sight problems – where blood vessels in the retina are involved.

People with type 2 diabetes need to take regular, moderate exercise to control their blood sugar levels. They have to eat carefully and plan their diet, so that sugar is released steadily.

All diabetics should keep a sweet snack or sugary sports drink at hand when they exercise in case their blood sugar level falls too low.

Type 1 diabetes can be treated with daily insulin injections and careful diet.

Which of these items contain sugars that are quickly absorbed? Which contain complex carbohydrates that release sugars slowly when digested?

In the 1920s scientist Frederick Banting tested different pancreas extracts on diabetic dogs. The purified active ingredient – insulin – was remarkably effective on patients, some of whom were close to death.

Questions

1 Suggest why insulin has to be injected – why can't it be swallowed like other medicines?

2 Make a table to compare the causes, symptoms, and treatments for type 1 and type 2 diabetes.

3 Outline how someone could reduce their risk of developing type 2 diabetes.

4 Your classmate feels faint after PE. You notice that she is wearing a wrist band showing that she is diabetic. You rush to tell your teacher. What do you think might have happened? What would you suggest that they do to help?

Key words

✓ **pancreas**
✓ **hormone**
✓ **insulin**
✓ **diabetes type 1**
✓ **diabetes type 2**
✓ **obese**

Find out about

- ✓ how exercise can help you maintain a healthy body mass

Why is exercise part of a healthy lifestyle?

Everyone knows that exercise is important for staying healthy. When you exercise you strengthen your muscles, improve your co-ordination, and develop your self-discipline. You also use up energy that you have taken in as food. If the energy that you use up matches the energy that you take in, then your body mass will stay fairly constant.

We all know that we need to exercise to stay in shape.

There are lots of ways to enjoy keeping fit.

How much energy does exercise use up?

Doing different activities will use different amounts of energy. The longer the time in a day that you are active, the more energy you will expend. You should always think carefully before starting a strenuous new exercise regime and ask your doctor if you are unsure.

Builders' breakfasts are highly calorific. Their intake matches the strenuous physical work that they do so their mass remains constant.

Activity	Energy (kJ/kg/h)
Sitting quietly	1.7
Writing	1.7
Standing relaxed	2.1
Vacuuming	11.3
Walking rapidly	14.2
Running	29.3
Swimming (4km/hour)	33

Energy used in different activities.

Sir Ranulph Fiennes pulled his own sledge unaided across the Antarctic continent. Although he consumed about 12,000 calories a day he still lost body mass.

What is a healthy body mass?

Doctors have studied the links between body mass and illness. If your mass is too great for your height then you may be at an increased risk of illness. It can also be dangerous to have a mass that is too low.

Exercise will not only help you to maintain a healthy body mass but can also bring you a lot of enjoyment and rewards. You might build up a circle of friends around your activities and be better placed to cope with the stresses of everyday life. You will get into good habits that will help to keep you healthy for your whole life.

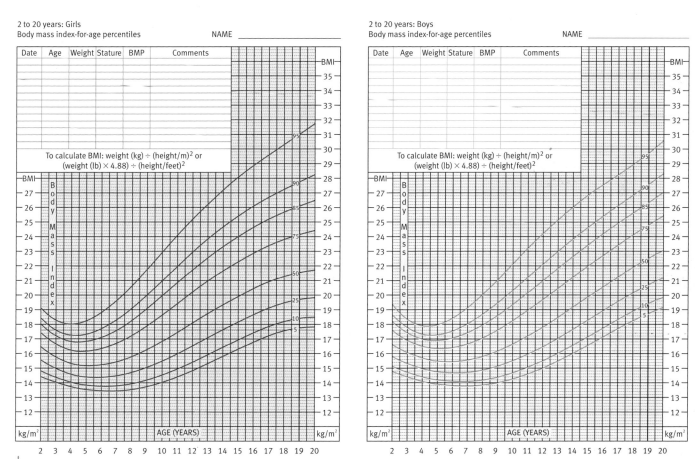

Graphs showing healthy height and body mass for young people in the UK. The numbers on the lines show the percentage of people with each BMI.

Questions

1 Find out how much energy (kilojoules) your favourite snack contains. Describe how much exercise you would need to do to 'work off' this energy.

2 Exercise can help to keep your body in shape. Describe the other benefits that doing exercise might bring.

Find out about

- ✓ how a poor diet could increase your chances of illness

These starving refugees have not had enough food.

During the Second World War food was in short supply in the UK. Food was rationed: there were strict controls on how much food everyone received. Despite this many people's diet was healthier than it is now – can you think why?

You are what you eat

In 2008 the United Nations reported that nearly one billion of the six and a half billion people on earth are starving. For economic, political, and environmental reasons (such as wars and natural disasters), they do not get enough to eat. But for most people in the UK food is abundant. Food can be flown in so we can eat fruit and vegetables that are out of season here. Modern farming methods and commercial food production have driven down prices and increased choice. It is still important to pay attention to the amount and type of food you eat to stay healthy.

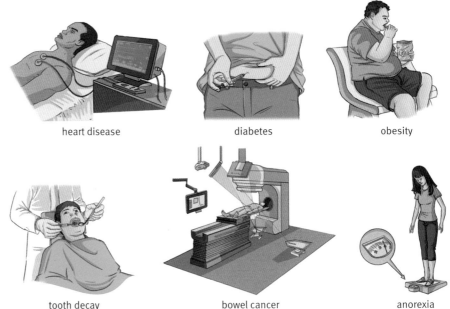

heart disease   diabetes   obesity

tooth decay   bowel cancer   anorexia

An unhealthy diet can lead to serious illness.

Risk factors

Things that increase our chance of becoming ill are called 'risk factors'. There are many lifestyle choices that we can take that have an impact on how healthy we are. Having an unhealthy diet increases the risk of you becoming ill – even if it takes several years before the symptoms appear. Scientists examine very large samples of the population to find links between diet and health as the patterns can be complex.

Looking for patterns

Scientists know that people in different countries have different rates of illness. Some of this can be explained by differences in the foods that people eat. For example, people in Japan eat less red meat and more fish than people tend to in the UK. As a result, Japanese rates of **heart disease** are lower than in the UK.

People in developed countries have a higher rate of bowel **cancer** than people in developing countries. Diets in developing countries are usually high in fibre from vegetables so food passes quickly through the digestive system. Refined, processed foods travel more slowly through our intestines so any toxins are in contact for longer.

Look at the following information about diet and cancer – will it change what you choose to eat?

Factor	Increased risk of cancer	Decreased risk of cancer
Alcohol	Breast, larynx, oesophagus, pharynx, oral cavity, liver and bowel	
Fruit and vegetables		Larynx, oesophagus, pharynx, oral cavity, stomach and lung
Processed and red meat	Bowel	
Fibre		Bowel
Salt	Stomach	
Dairy products		Bowel

Some foods increase the risk of cancer but some reduce it.

Questions

1 Describe what you understand as an 'unhealthy diet'.

2 Which foods should you eat only in moderation to stay healthy? Explain why.

3 Which foods should you try to include in your diet to stay healthy? Explain why.

4 Your friend suggests you should both get chips and a burger at the 'Greasy Spoon Cafe'. Make a list of replies that would lead to a healthier option for you both. Remember, your answers need to be practical!

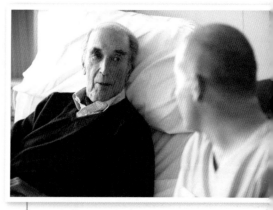

The Cancer Research UK EPIC programme is part of a project to monitor the health of 100,000 people across ten European countries. This large data sample, gathered over many years, will help scientists to find links between diet and illness.

Key words
✓ **heart disease**
✓ **cancer**

We can get cheap, 'tasty' food whenever we want – is this a good thing?

Topic B7.4: Learning from nature

Stone statues (Moai) of Easter Island in an empty landscape.

Why study natural ecosystems?

Are we destroying our environment? Can the planet continue to support its growing human population? We are part of nature, but we are using natural resources too quickly. By studying nature we can learn to live in a sustainable way.

The Science

In nature, nutrients, gases, and food pass from species to species in a cycle. Very little is wasted so there is no pollution. Humans have changed this closed loop. We use natural resources to manufacture products such as paper, computers, and cars. We alter ecosystems, destroy wildlife, and produce a lot of waste. We are damaging the life support systems we depend on.

Ideas about Science

Science has provided new technologies that help us to grow more food and harvest more fish and timber. These have enhanced the quality of life for many people but have also had harmful effects on the environment. We must find alternative, sustainable methods before too much damage is done. Decisions about the exploitation of wildlife should be guided by ethics and international regulation.

A variety of organisms in their ecosystems.

Find out about

- ✔ **linear and closed-loop systems**
- ✔ **natural ecosystems as closed-loop systems**

Key words

- ✔ **linear system**
- ✔ **sustainable**
- ✔ **closed-loop system**
- ✔ **microorganisms**
- ✔ **ecosystem**

Questions

1 Explain in your own words what you understand by the term 'sustainable'.

2 Explain why linear systems for making products end up with toxic waste and depletion of natural resources.

3 Suggest how a manufacturing system can become more sustainable by copying natural closed-loop systems.

4 Give two reasons why our present way of life is not sustainable.

5 Explain why oxygen is taken up by some organisms and lost by others, in a natural ecosystem.

Linear systems

Easter Island was fertile and full of trees, but the people who lived there cut down the forest without ensuring that new trees grew. Look at the photo on page 228 to see what it is like now.

How can we avoid the fate of the people of Easter Island? Most of us live in a 'take–make–dump' society. We *take* natural resources from our environment, *make* them into products, and then *dump* the waste. This **linear system** is not **sustainable**. It can only continue for a short time before things go wrong because:

- fossil fuels, such as oil, are running out
- natural resources such as timber, clean air, fresh water, fertile soil, and fish stocks are being used more quickly than they are being replaced
- making products uses a lot of energy from fossil fuels, and creates a lot of waste
- waste also comes from broken and worn-out products that we throw away
- waste can be harmful to people and wildlife, and can stay in the environment for a long time
- waste means rare resources such as metals are spread thinly around the environment, so they can't be reused easily.

Natural closed-loop systems

Nature works in **closed-loop systems**. A closed-loop system has no waste. Output from one part of the system becomes the input for another part, so X's waste becomes Y's food. For example, plant waste, such as dead leaves, is eaten by snails. Faeces from snails is broken down by **microorganisms** (bacteria and fungi). Microorganisms release nitrogen and phosphorus from the faeces. Nitrogen and phosphorus are taken up again by plants. Chemicals go around in a cycle and nothing is wasted.

Natural closed-loop systems are called **ecosystems**. Examples of ecosystems include lakes, woodlands, grasslands, beaches, and coral reefs. In all of these ecosystems there is a community of organisms interacting with the non-living elements of their habitat.

Each species has its own job in the ecosystem. It could be a plant, herbivore, carnivore, decomposer, or parasite. Every species uses different foods and produces different products. All depend on each other.

Organisms in an ecosystem exchange materials with their surroundings. Plants take up carbon dioxide for photosynthesis, and animals take in food for growth and reproduction. At the same time organisms return substances, such as oxygen from photosynthesis, carbon dioxide from respiration, and faeces from digestion. The ecosystem is a closed loop for materials.

You can't recycle energy, so energy is always part of a linear system. This is why closed-loop systems need a constant supply of energy from a sustainable source.

What can we learn from natural ecosystems?

Can we change the way we live, from a linear 'take–make–dump' system, to a closed-loop system? A closed-loop lifestyle can be described as 'take–reuse–recycle'. Can copying nature provide us with all the answers we need for sustainable living?

Linear system

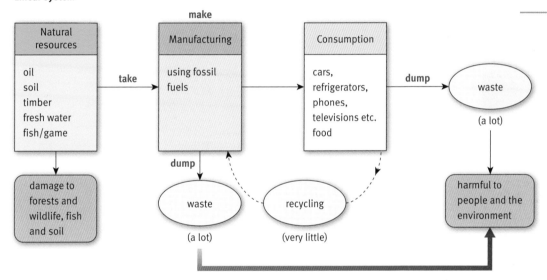

In a linear system, waste products can build up to levels where they become toxic to humans and wildlife. Closed loops reuse and recycle, minimising waste.

Closed–loop system

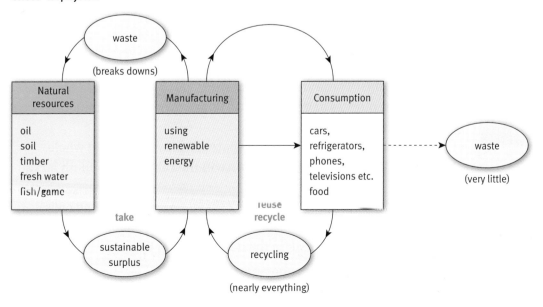

Find out about

- ✔ **how waste materials are reused in ecosystems**
- ✔ **human societies that live in a closed-loop system**

Burying beetles on a dead animal's body.

Barn owls in their nest.

Key word
- ✔ **reactants**

In natural ecosystems, waste materials from one species are often used as food by another species. Waste materials can also be used in chemical reactions as **reactants** by another species. Reactants include nitrate and phosphate waste from microorganisms. These reactants are taken up as plant nutrients.

Closed-loop lifestyles

Barn owls

Barn owls live all over the world. You may have seen one, hunting along the roadside for voles. All the inputs required by a barn owl come from its habitat, and all its waste is used by other organisms in its habitat – it lives in a closed-loop system.

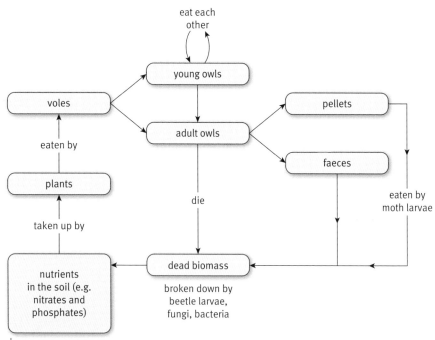

A barn owl lives in a closed-loop system. There is nothing left over and no pollution.

- **Droppings:** These are broken down by microorganisms. This releases nitrates and phosphates. Plants take up these minerals for growth.
- **Pellets:** These are indigestible parts of prey, swallowed whole. Bones from pellets slowly dissolve in the rain. Clothes moth larvae eat the fur.
- **Dead owls:** Burying beetles lay eggs in the bodies of dead owls and the larvae eat the dead flesh. After a few weeks very little is left. If a baby owl gets weak in the nest it is eaten by one of its brothers or sisters.

Maasai people

A few human communities, such as the Maasai, still live in a closed-loop, sustainable way.

Maasai live on the grasslands of Kenya and Tanzania in East Africa. They move from place to place, finding fresh grass for their herds of cattle, sheep, and goats. They make temporary villages. Their round huts are made from sticks, grass, mud, ash, and dung. The village is surrounded with a fence made of thorny branches. At night-time the domestic animals are brought inside for safety.

Maasai do not hunt game or birds, and they do not eat much meat. They drink milk and also take blood from the jugular vein of cattle as a ritual or as medicine. They use some wild plants as food. When they die, their bodies are left out for scavengers.

A Maasai temporary hut is fully biodegradable when the community moves on.

Questions

1 Explain why the Maasai diet causes little damage to their environment.

2 Traditionally Maasai used to wear clothes made of animal skins. More recently they have bought cotton clothes. Explain which of these two types of clothing is more sustainable.

3 Should the governments of Kenya and Tanzania encourage Maasai people to live in permanent villages? Suggest two reasons for and two against.

4 Explain whether you think it would it be possible for all of us to live like the Maasai.

Find out about

- types of waste product in natural ecosystems
- the way waste products become food or reactants for other organisms
- storage and movement of chemicals through ecosystems

Cycling of materials

You have probably shuffled through piles of leaves and fallen acorns or apples on the pavement or on a woodland path in autumn. In spring the leaves may be replaced by carpets of cherry blossom. Most of this fallen material disappears quite quickly.

Fallen branches, leaves, petals, and fruits are called **dead organic matter** (though seeds inside the fruits can still be alive). Dead organic matter (DOM) is any material that was once part of a living organism. It also includes waste material from animals, such as faeces and bodies.

When DOM falls to the ground there are many different organisms waiting for it.

- Worms nibble leaves and grind them up in their gut with soil. The leaves are then digested by enzymes.
- Threads of fungi in the soil release **digestive enzymes** that break down DOM.
- Dung beetles roll up faeces into pellets, then bury the pellets and lay eggs in them. There are many different dung beetle species, using different types of dung.
- Hundreds of different kinds of bacteria live in the soil. They make a great variety of enzymes. Without bacteria there would be no recycling of reactants such as carbon and nitrogen in ecosystems.

Fallen cherry blossom is not waste in a natural system.

Dung beetle on rabbit droppings.

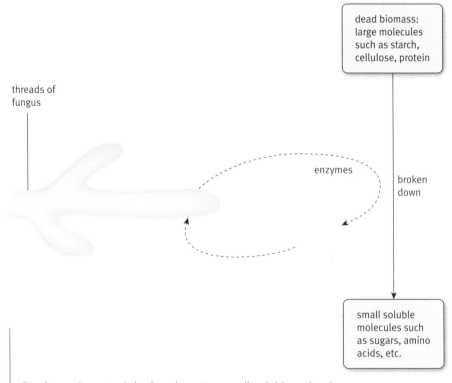

threads of fungus

dead biomass: large molecules such as starch, cellulose, protein

enzymes

broken down

small soluble molecules such as sugars, amino acids, etc.

Dead organic matter is broken down into small, soluble molecules, which are absorbed by fungal threads and used for growth.

Bacteria and nitrogen

Nitrogen gas is all around us in the atmosphere. Nitrogen is also a component of protein and DNA, so all organisms – including humans – contain nitrogen. Several different types of bacteria are essential in the nitrogen cycle.

- Nitrogen-fixing bacteria take nitrogen from the air and make it into nitrates. These bacteria live freely in the soil or inside the roots of some plants, such as beans and clover. Plants then use the nitrates to make protein and DNA.
- One type of decomposing bacteria in the soil breaks down proteins and amino acids into ammonium ions.
- Other bacteria in the soil convert ammonium ions into nitrates. Plants then take up the nitrates through their roots.
- In **anaerobic** soil, bacteria change nitrates into nitrogen gas. Anaerobic soils have no oxygen and include soils that are waterlogged or compacted.

Key words

- ✓ **dead organic matter**
- ✓ **digestive enzyme**
- ✓ **anaerobic**

Questions

1 Describe how a carbon atom in the carbon dioxide in the air can become a carbon atom in an animal.

2 Describe how a nitrogen atom in an animal can become a nitrogen atom in the air.

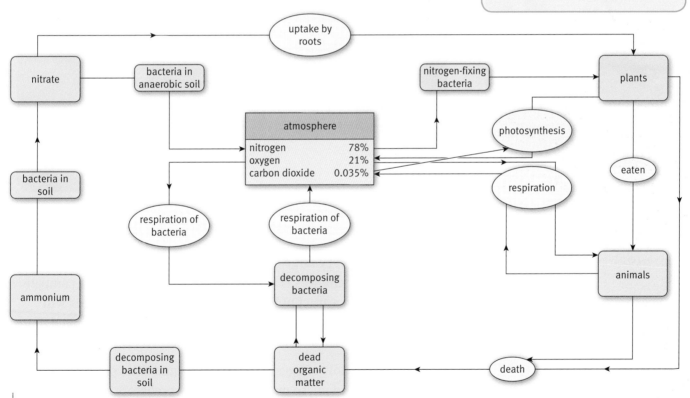

Exchange of reactants in an ecosystem.

Bacteria and carbon

Bacteria in the soil also break down carbohydrates in DOM, such as cellulose and starch. The carbohydrates provide the bacteria with glucose for respiration. Respiration releases carbon dioxide into the atmosphere. The carbon was originally taken from the atmosphere by photosynthesis. Carbon is a reactant that is recycled.

Find out about

- ✔ **the way plants and animals improve their chances of successful reproduction**
- ✔ **the way waste materials are recycled in ecosystems**

Abundance without waste

If you suffer from hayfever you will know that plants, especially grasses, make a lot of **pollen**. The pollen grains blow in the wind to other grass flowers. This transfer of pollen is called **pollination**. Pollen contains the male sex cells of flowering plants. Pollination is needed for **flowers** to fertilise the female sex cells and produce seeds.

Producing lots of pollen increases the chance that some will reach other flowers and pollinate them. This is a natural insurance policy.

For an oak forest to survive, each tree must produce just one acorn that grows into a mature tree in its lifetime. Most acorns do not survive, but their materials are recycled into the ecosystem.

Colourful flowers are pollinated by insects, such as bumblebees. Bees carry pollen on their furry bodies and in pollen sacs attached to their legs.

Plants also make very large quantities of petals and **fruit**. They are maximising their chances of reproducing successfully. Plants compete with each other to attract **pollinators**. Lots of big bright petals will attract many bees and other insects. When the flower is pollinated the petals have no more use, and fall to the ground.

Animals produce lots of eggs and very large quantities of sperm. Eggs contain a lot of protein, so many are eaten by predators before the eggs hatch. Young hatchlings are also eaten in large quantities. Birds' eggs are fertilised within the body and mammals' eggs develop inside the body as well. This strategy is safer, so an insurance policy is not so important. For this reason, birds and mammals make fewer eggs than fish or insects.

Laying hundreds of eggs is a strategy for successful reproduction in some species such as this squid.

A male producing many sperm cells increases his chances of fathering offspring. Many sperm cells die or do not find the egg. This is a male's strategy for successful reproduction.

○ Many eggs are taken by predators such as this monitor lizard.

This ecosystem is not a perfect closed loop because some migrating ○ birds do not return.

Is nature wasteful?

Branches and leaves fall off trees. Large quantities of petals, fruits, eggs, and sperm are produced but not used. Fungi make millions of spores, but most spores do not grow into new fungi. In a **stable ecosystem** all this DOM decomposes, and the materials are recycled. It is very close to being a closed-loop system, so very little is wasted. Water, nitrogen, carbon, and oxygen simply circulate around the ecosystem.

Natural systems do not need to worry about wasting energy because they have a constant source – the Sun.

Are ecosystems perfect closed loops?

The organisms in an ecosystem constantly recycle food and reactants in a closed-loop system. But many ecosystems have inputs and outputs – they are not perfect closed loops.

- When birds migrate in the winter, not all of the birds will return in the spring. So some material (their bodies) is permanently removed from their summer homes.
- Rivers carry branches, leaves, nutrients, and silt away and deposit them further downstream.
- Coral reefs grow larger as they take minerals from the water.

A stable ecosystem has no overall gains or losses: the inputs and outputs are balanced.

Key words

- ✓ pollen
- ✓ pollination
- ✓ flowers
- ✓ fruit
- ✓ pollinators
- ✓ stable ecosystem

Questions

1 Explain why very few seeds result in new plants.

2 Suggest why mice and robins produce fewer eggs than frogs or fish.

3 Give two ways (other than those mentioned here) that materials can move into or out of an ecosystem.

4 Draw a flow diagram to show how natural systems can produce more material than is needed and still be sustainable.

5 Explain how most human manufacturing systems are different to natural systems, making them unsustainable.

Find out about

- ✓ **ecosystem services provided by forests**
- ✓ **ways to manage and conserve water supplies and soil**

Most of us take it for granted that the Earth will continue to provide us with fresh water to drink, clean air to breathe, a fertile soil for growing crops, and a supply of food such as fish and game. **Ecosystem services** include the different ways that living systems provide for human needs.

The growing human population is putting increasing pressure on ecosystem services worldwide. Understanding how ecosystem services work helps us to get the best out of them. This should help to avoid disasters similar to Easter Island, where ecosystem services broke down.

Clean water

Mexico City has a population of around 20 million people. Rainfall comes between June and September and mostly falls on the surrounding mountains. There is a danger of flooding during the rains, but severe water shortages occur in the dry season. Mexico City has damaged the natural closed-loop water cycle by **deforestation** (cutting down the forest) and cattle grazing. Each year more water is used in the city than is replaced by rainfall.

Mexico City.

The forests above Mexico City provide a natural ecosystem service.
- The soil is rich in DOM from rotting leaves, and holds water like a sponge. The water gradually drains into the rivers supplying the city, so that there are reserves for the dry period.
- Tree roots reduce **soil erosion** by holding the soil together and leafy branches prevent rain falling directly onto the soil. Rain drips

slowly off the trees and is absorbed rather than running off the soil surface. Soil is not washed away. Eroded soil would silt up rivers and block drains.

• Water evaporation from the forest canopy generates clouds and rain, and cools the air.

Now the forests above Mexico City are being protected and restored. The people of Mexico City have recognised that breaking a natural closed-loop system destroys a vital ecosystem service. The city's water policy includes:

• protecting forests on the slopes of Mount Popocatepetl above the city
• diverting flood water into wells to restore underground water levels.

Fertile soil

The Aran Islands off the west coast of Ireland are mostly covered in bare rock. People fled there from Cromwell in the 17th century. There was hardly any soil for growing crops. They made their own soil by using layers of sand and seaweed mixed with animal dung.

Bare rock only occurs where there is extreme wind, wave action, drought or ice. Everywhere else, rock is usually covered by a thick layer of fertile soil. Soil is formed from broken down rocks and DOM.

Earthworms play an important role in breaking down DOM, and mixing and aerating the soil. Ploughing damages earthworms. Soil then becomes compacted and crops do not grow so well. **Direct drilling** is now being tested. For example, rape seed can be planted directly into wheat stubble (stalks of newly cut wheat) with no ploughing. The worms are not damaged and the soil is more fertile. A good crop of oilseed rape is produced. This shows how we can grow crops without destroying the ecosystem services we depend on.

> **Key words**
> ✔ **ecosystem services**
> ✔ **deforestation**
> ✔ **soil erosion**
> ✔ **direct drilling**

Direct drilling protects worms and keeps the soil healthy.

> ## Questions
>
> 1 Describe how soil is formed.
>
> 2 Explain how forests help to keep rivers flowing during the dry season and suggest how they prevent flooding.

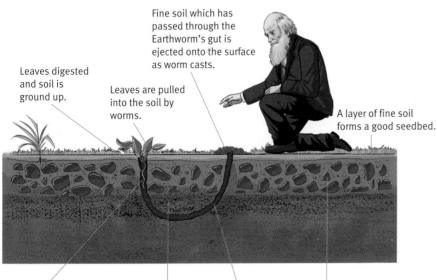

Fine soil which has passed through the Earthworm's gut is ejected onto the surface as worm casts.

Leaves digested and soil is ground up.

Leaves are pulled into the soil by worms.

A layer of fine soil forms a good seedbed.

Air gets into soil through earthworm's burrows.

More DOM in soil for plant nutrients.

Burrows help water drain easily.

Stones are buried by the action of Earthworms.

Much of the hard work of making soil is done by earthworms. They provide an essential ecosystem service. This was first described by Charles Darwin.

The end result of our take–make–dump lifestyle.

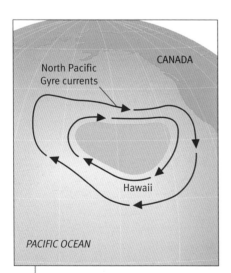

The Great Pacific gyre currents swirl around in the north Pacific, carrying plastic and other debris that harms marine life.

Our take–make–dump linear lifestyle produces a lot of waste that we need to get rid of. Waste comes from households, industry, and burning fossil fuels.

Most of our waste can be decomposed by microorganisms – but not all of it. Microorganisms are unable to produce the right enzymes to break down some waste materials. These substances are called **non-biodegradable**. For example, glass, synthetic fabrics, some pesticides, and many plastics are non-biodegradable. This means that they stay around in the environment for a very long time.

Plastics in the sea

Plastic floats down rivers and enters the sea. It can be eaten by sea birds, fish, and turtles. They cannot digest the plastic, so their guts get blocked. Wave action breaks some plastics down into fine granules. Small animals filter plastic granules from the water instead of their normal food. These small animals are the food supply of fish.

Bioaccumulation of toxic chemicals

Shiny coloured paper from magazines and printed cardboard contains heavy metals. Bleached paper, cardboard products, and certain plastics contain chlorine. Chemicals called **dioxins** are made when these bleached products are burned with other waste. Dioxins, heavy metals, and other chemicals from human waste accumulate in ecosystems. **Heavy metals** have been linked to birth defects and cancer in humans. Exposure to dioxins is associated with cancer, birth defects, and problems with the immune system.

Bioaccumulation

Chemicals that are released in small quantities can build up to toxic concentrations through food chains. This is called **bioaccumulation**. For example, a vole might take in a small amount of pesticide in its food (grass). The pesticide is stored in the vole's body as it cannot be broken down. When an owl eats a vole, all the pesticide goes into the owl. Each vole adds a bit more pesticide to the owl, as the owl also can't break down pesticides. Eventually enough pesticide accumulates in the owl to make it infertile, or even to kill it.

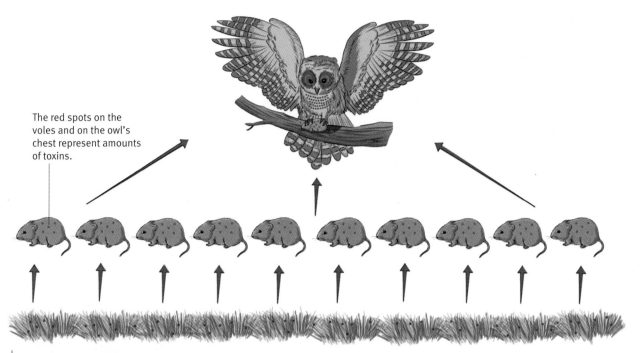

The red spots on the voles and on the owl's chest represent amounts of toxins.

Chemicals that are not broken down in living organisms can **bioaccumulate**. Toxic levels of heavy metals, pesticides, and other chemicals build up in birds, fish, and potentially humans. These chemicals are by-products of unsustainable human systems that produce waste.

Heavy metals and dioxins that are washed into rivers and the sea are concentrated in plankton, then fish and sea birds. If humans eat fish that have accumulated metals such as mercury, the levels can be high enough to cause concern.

What can we learn from natural waste disposal?

In natural ecosystems all waste is broken down by microorganisms, using enzymes. The substances released are then used by other organisms. Waste does not build up to toxic levels in natural systems.

We should try to imitate this process by making waste **biodegradable**. We can use cardboard instead of polystyrene for drink and food containers. Starch pellets can be used for packaging instead of plastic bubble wrap. Products and materials should be designed with a view to what will happen to them at the end of their useful life.

Questions

1   Turtles mistake plastic bags for jellyfish. Explain why turtles cannot digest plastic bags.

2   Explain how granules of plastic in the sea might affect fish stocks.

3   Explain why starch pellets are better than plastic for packaging.

Key words

- ✓ **non-biodegradable**
- ✓ **dioxins**
- ✓ **heavy metals**
- ✓ **bioaccumulation**
- ✓ **biodegradable**

Biological waste as part of a closed loop

When humans remove **biomass** from an ecosystem, it can cause irreversible changes. Biomass is any biological substance we harvest, such as grass, crops, wood, fish, and game. Agriculture involves removing biomass from fields. Soil nutrients need to be replaced using fertilisers.

Natural systems use waste as food for other organisms. To make agriculture into a closed-loop system, human faeces and urine would need to be returned to the fields as fertiliser. By collecting and removing human biological waste from the system, the system becomes linear, take–make–dump.

Cow dung goes some way towards replacing the nutrients lost through the cows eating the grass.

Fertiliser	500 litres of urine	500 litres of faeces	Total	Fertiliser needs of 250 kilograms of cereal
Nitrogen	5.6 kg	0.09 kg	5.7 kg	5.6 kg
Phosphorus	0.4 kg	0.19 kg	0.6 kg	0.7 kg
Potassium	1.0 kg	0.17 kg	1.2 kg	1.2 kg

The typical amount of nitrogen, phosphorus, and potassium in human waste. This can be compared with the fertiliser needs of a cereal crop.

Some societies process human waste to use as fertiliser. But this risks introducing high levels of toxins into food crops by **bioaccumulation**. It could also transmit infectious diseases. Other biodegradable organic wastes such as animal manure, unwanted food, and plant waste can be used to help make agriculture a closed-loop system.

Key word

- ✔ **biomass**

Eutrophication

Using organic fertiliser may reduce harm to the local environment. Nutrients from non-organic fertilisers often wash off fields and into rivers and lakes. Faeces and uneaten food from fish farms also add nutrients to water. These nutrients cause **algae** (simple green water plants) to grow rapidly. The water goes green in an **algal bloom**. The algae soon die and decay in the water. Bacteria causing the decay take dissolved oxygen from the water for respiration. The oxygen levels in the water go down quickly. This can kill animals in the water, such as fish. It also kills aquatic plants that would normally add oxygen to the water as they photosynthesise. This is called **eutrophication**.

The River Thames is now clean. It has been restocked with salmon from hatcheries to allow sustainable harvesting.

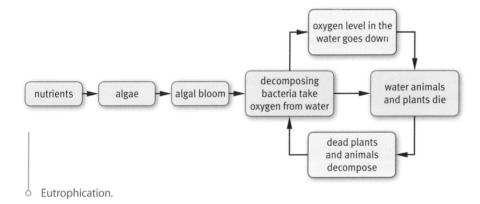

Eutrophication.

To make the agriculture system more 'closed' and so more sustainable, farmers can use manure and other organic fertilisers. These break down more slowly. Nutrients are released at the rate the crop can absorb them, so they do not get washed away in the rain. **Crop rotation** uses plants, such as clover, with **nitrogen-fixing bacteria** in their roots. Leaving a field with a **fallow crop** of clover once every few years replenishes the soil with nitrogen compounds.

Questions

1 Discuss whether agriculture to produce human food can be fully sustainable.

2 Suggest why, in some parts of China, visitors would traditionally be invited to leave behind their faeces before departing from the area.

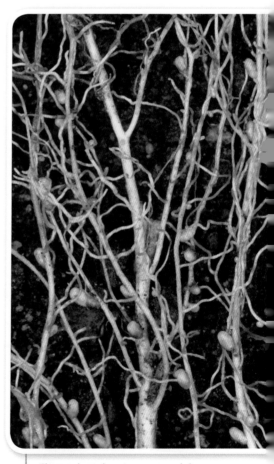

Clover plant showing root nodules. Root nodules of clover plants contain nitrogen-fixing bacteria. The bacteria and clover plant make nitrogen compounds by fixing nitrogen in the air. Clover is then ploughed into the soil to act as a green fertiliser.

California's almonds

Brazil nuts come from wild trees in the rainforest. The trees make a surplus of nuts, so there are plenty to spare. Many nuts are eaten by wild animals. The natural closed-loop system is not greatly changed.

California produces 80% of the world's almonds. Tonnes of almonds are taken away, creating a linear system. Not enough nutrients return to the ground. Fertilisers have to be added. The crop is low in **biodiversity** because there are no other crop species or weeds to support wildlife.

Pest insects and fungi have few natural enemies in an almond crop. They spread quickly and do a lot of damage. The pesticides used on California's almond crops kill pollinating insects, such as bees, in addition to the insect pests.

Over a million bee hives are transported to the Californian orchards every year. The honeybees do their work on the almond blossom in about three weeks. The hives are then moved to orange and apple crops. The bees are kept very busy.

California's almonds are an example of **intensive agriculture**. Intensive agriculture is linear, as it uses a lot of inputs, such as fertilisers and pesticides. Intensive methods can grow a lot of food, but crop failures are more likely. In the long term they are unsustainable.

Question

3   Suggest two ways California's almond industry could become more sustainable, using the idea of closed-loop systems.

Just almond trees are grown in over an area of around 243,000 hectares in California.

Unsustainable fishing

When you buy fish and chips, do you ask where the fish came from? North Sea cod stocks have crashed, but Icelandic cod is well managed. Consumers have a choice and can have a powerful influence.

Many of the world's fish stocks are overfished. Not enough fish are left to breed and replace the fish caught. But why should a fishing boat reduce its catch for everyone else's benefit? Other boats will simply take the fish instead.

Solutions to overfishing

The European Community sets **quotas** for North Sea cod. Countries agree the total fish catch that a boat can make each year. This does not solve all the problems.

- Politicians want to protect jobs and please electors. They often set higher quotas than scientists recommend.
- Fishing has to be monitored and policed.
- If a boat catches too many fish or the wrong species, they are often dumped.
- Fishing fleets move to unprotected areas, such as the African coast. Then the local people may resort to piracy or eating bush-meat because their fish stocks are reduced.
- There is more fishing for deep-sea fish, which grow very slowly. Deep trawling also damages the sea bed.

Fishing bans

The Mediterranean bluefin tuna fishery is close to collapse due to unsustainable modern fishing methods being used. In February 2010, the European Union proposed a fishing ban for bluefin tuna.

The collapse of predatory tuna fish may allow other **predators**, such as squid, to move in. If a ban is imposed too late, tuna populations may not recover. Their position on the food web will have been taken. Newfoundland's cod fishery collapsed in 1992 and is still unfishable. Protection came too late.

Fish farming (aquaculture)

Salmon farms produce cheap salmon, but are not sustainable. Food must be added and dirty water taken away. This linear system can cause environmental damage. Overfishing anchovies to provide food for salmon and polluting the local environment are just two of the problems.

Tuna fishing.

Questions

4 Suggest why politicians often set fish quotas higher than scientists would like.

5 The orange roughy is a deep-sea fish that grows very slowly. Suggest why it is not sold by some supermarkets.

6 Explain why some fish stocks do not recover after they have been overfished.

Find out about

- ✓ **the effects of replacing natural vegetation with farmland**
- ✓ **how farmland can turn to desert**
- ✓ **ways to reduce the effects of farming on biodiversity and soil**

One fifth of the world's population lives where the rainfall is less than 400 millilitres a year. These dry-land ecosystems are easily damaged and can turn to **desert**. A desert is a place where no crops can grow. What can we learn about sustainable agriculture from people who live in dry lands?

Desertification in the Sahel

The Sahel dry-land zone lies south of the Sahara. Local people have developed many techniques to prevent the land turning to desert, a process called **desertification**.

- Herds are moved from place to place to prevent overgrazing.
- Trees are not cut down in fragile areas. Acacia trees are planted as windbreaks.
- Thorny branches and rocks are used to reduce erosion in stream beds.
- Hundreds of small pits are dug in fields in the dry season. Compost (decaying plant material) is placed in the pits and covered with soil. Termites and fungi live in the compost and make the soil fertile. When the rains arrive, water drains into the pits, making a seedbed for food crops such as sorghum and millet.

In the 1970s and 1980s there were severe droughts in the Sahel. More than 100 000 people starved to death. International aid saved a lot of lives, but now a long-term solution is needed.

Today more Sahel people live in settled communities. The population has risen, though the area is much less crowded than western Europe. The rains have been better and people have been tempted to use intensive, linear-system farming methods, using fertiliser and pesticides. Some of the traditional skills have been lost. Much of the natural vegetation has been cleared, and there is **overgrazing**. Goats and other animals have eaten nearly all the plant cover. Wind and rain erode the soils.

Natural vegetation has been lost. The desert advances. Soil is lost and rivers silt up, increasing the risk of flooding when the rains do come.

Collection of native seeds for Kew Millennium Seed Bank in Burkina Faso.

The United Nations has a 'Desert Margins Programme for Africa'. Their aim is to encourage the best farming methods, and to conserve the unique native desert plants and animals. **Native species** are adapted to the dry conditions. They grow much better than imported species. Kew Royal Botanic Gardens is collecting seeds of native plants in Burkina Faso, with the help of local people. The seeds are being used for desert restoration.

Locust bean tree *Parkia biglobosa* provides protein-rich seeds, fuel, and medicine. It also fixes nitrogen.

Shea tree *Vitellaria paradoxa* provides nuts and oil.

Climate change and population pressures continue to threaten the survival of the Sahel's people. A return to traditional closed-loop systems, using native species and crop varieties, could bring sustainable solutions.

Solutions based on natural systems are also needed in developed countries. Often trees and shrubs are removed to plant crops in large fields. This is convenient for large farm machinery. Removal of hedges and trees has led to soil erosion, reduction of pollinating insects, and reduction of the natural predators of crop pests.

Questions

1 Explain the word 'desertification'.

2 Look back to Section E. Use the information to explain why planting and looking after trees helps to prevent soil erosion and desertification.

3 Explain why it is better to grow native desert plants than introduced ones.

4 Suggest why farmers should be encouraged to plant hedges to divide up their fields of crops.

Gamba grass *Andropogon gayanus* can survive fire, restores grassland, and is used for thatch.

Find out about

- the benefits of conserving natural ecosystems
- the value of wild animals to human welfare
- ethical reasons for conserving wildlife
- tensions between conservation and local people

Year	Tiger numbers in India
1800s	45 000
1972	1827
1990s	3500
2010	1500

Biodiversity is rapidly declining. 'Biodiversity' means the variety of life – the number of species and variation within each species. There are often tensions between conserving natural ecosystems and the needs of local people. In India, rapid economic development and a growing population are putting pressure on tiger reserves. Conserving ecosystems is beneficial to humans, as it protects the ecosystem services they provide.

The price of a tiger

Is a tiger worth more dead or alive? Sadly, living tigers are undervalued, but there is a huge market for dead tiger parts for traditional medicines. India is losing its tigers very rapidly. They may soon be extinct in the wild.

In 1972 the Indian Government established 28 tiger reserves spread across India. People were moved away and the vegetation was allowed to recover. Game increased rapidly and tiger numbers rose to 3500 by the 1990s.

In 1993, China banned the sale of farmed tiger products. More wild tigers were poached for traditional medicines. By 2008 tiger numbers in India declined to just 1400.

The greatest good

Should people be forced to move away from tiger reserves? The buffer zones around the reserves make good grazing land for cattle and a source of wild game. There is little other good land for people to go to. But moving people away could benefit more people than it harms.

Bengal tiger.

What are tigers good for?

Tigers are at the top of the food chain. A good tiger population means that the whole ecosystem is healthy. Tigers cannot survive without samba deer and chital prey. The deer will not survive without the right plants to eat.

Preserving India's ecosystems ensures that native plants and animals will survive. Wild plants provide medicines and food. Wild game can be harvested by humans for protein. Forests control climate and water resources.

Should we value tigers for themselves and not just as a commodity?

The domestic chicken came from India. What other potential food sources do India's ecosystems hold?

Can we live without tigers? Tigers bring inspiration to many people. Tourists come to see them. Tiger reserves offer jobs to rangers and scientists. Without tigers fewer tourists would come, and the whole ecosystem would change.

There are **ethical** reasons (non-scientific reasons to do with right and wrong) for conserving tigers and other wildlife. Many people believe that wildlife has a right to exist, for its own sake, without human interference. Some people think it is wrong to use tiger parts for medicines or to keep tigers in captivity.

Questions

1 Banning farmed tiger products sounds like a good idea. What have been the unintended impacts of this on the environment?

2 Explain whether you think people should be moved away from tiger reserves. Give two reasons for and two against.

3 List four ecosystem services provided by India's forests.

Key words
- ✓ **biodiversity**
- ✓ **ethical**

Tigers bring economic and environmental benefits to India.

How many different uses of wood can you see around you? Did the wood come from your local area or from overseas? To use our forests sustainably we must know where wood comes from, and how the forests are managed. Most of the world's forests are being used unsustainably. They are being cut down more quickly than they are growing.

Old growth or new growth?

Some forests have never been cut down. They are called **primary forests** or old growth forests. They are very rich in biodiversity. They have taken millions of years to evolve and are irreplaceable. These forests are being destroyed rapidly for the needs of local people. They cut down forest for timber, to provide grassland for cattle, and to grow palm oil, soya, and **biofuels**. Loss of forests causes soil erosion, mud slides, **silting of rivers**, flash floods, loss of cloud cover, and drought. It also takes away a sustainable source of timber.

Biomass

Taking timber from a forest removes biomass. When biomass is removed it changes the natural closed-loop to a linear system. Nutrients are taken away in the biomass. If a crop is harvested, inputs of fertiliser are needed to replace the nutrients removed. Sustainable use of timber means replacing the trees and nutrients as quickly as they are taken away.

This diagram shows an island divided between sustainable forest and deforested land. Compare the ecosystem services on either side.

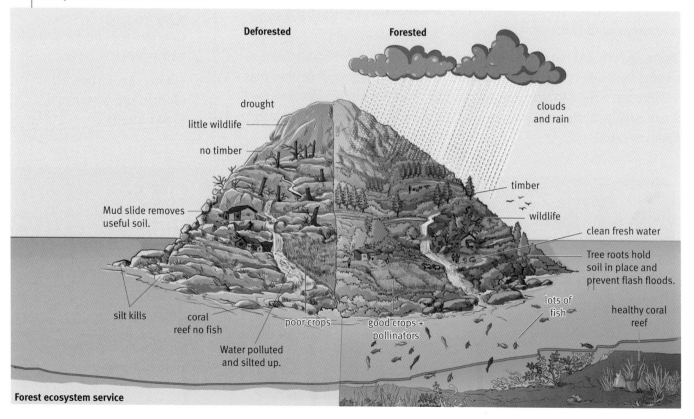

Deforested — Forested

drought
little wildlife
no timber
clouds and rain
Mud slide removes useful soil.
timber
wildlife
clean fresh water
Tree roots hold soil in place and prevent flash floods.
silt kills
coral reef no fish
poor crops
good crops + pollinators
lots of fish
healthy coral reef
Water polluted and silted up.

Forest ecosystem service

Where should we buy wood?

Many rich countries want to protect their remaining forests. They import a lot of their timber. This can damage forests in poorer countries.

One solution is to use **eco-labelling**. An eco-label shows that the timber was harvested from a sustainably managed forest. When we buy wood we can check the label to see where and how it was produced. Eco-labelling is still under development.

Restocking damaged forests

The National Forest in the East Midlands covers 518 000 hectares over three counties. The area was badly damaged in the past by coal mining and clay pits. Since the 1990s, forest cover has increased from 6% to 18% and more trees are being planted.

The growing forest removes carbon dioxide from the atmosphere by photosynthesis. Forest is a **carbon sink** as it stores carbon in the wood. It also helps to restore soils and cleans water supplies. The forest creates jobs through leisure and tourism. Wildlife spreads from areas of old woodland, increasing biodiversity. Local seed sources are used for the trees. This means the trees are adapted to local conditions. They will come into leaf and fruit at the right time for local birds and insects.

Conservation charities lease the Harapan Rainforest in Sumatra, to prevent it being used for logging. The forest contains tigers, elephants, hornbills, and the world's largest flower. As the tree canopy recovers, the humidity of the forest is increasing. The temperature varies less and fires do not spread so easily. The roots of the trees take up water, and this evaporates from the leaves. The water cycle of **cloud formation**, rainfall, and uptake by plants maintains the rainforest ecosystem.

Local people can harvest rattan, wild honey, and medicines sustainably from the recovered forest. This reduces tensions between local people and the conservationists. Local people get involved in the conservation effort. For example, young Sumatran volunteers visit the area and plant trees grown from local seed.

Key words
- ✓ **primary forests**
- ✓ **biofuels**
- ✓ **silting of rivers**
- ✓ **eco-labelling**
- ✓ **carbon sink**
- ✓ **cloud formation**

Questions

1 Many trees are imported from Eastern Europe for planting in Britain. Explain why this is not a good idea.

2 Explain why involving local people is important for conservation.

3 Explain how eco-labelling could help to protect forests.

What would happen if there were no elephants in the forest?

Find out about

- ✓ **the way oil forms**
- ✓ **why using oil is unsustainable**
- ✓ **oil as 'buried sunshine'**
- ✓ **the effects of burning fossil fuel on carbon dioxide levels**

We are living in the age of oil, but we must quickly move to a post-oil economy because:

- fossil fuels are running out
- burning fossil fuels is increasing carbon dioxide levels in the atmosphere.

Where does oil come from?

Oil comes from the dead bodies of minute plants and animals. They lived in the sea millions of years ago. They fell to the sea bed and were slowly covered by layers of sand and silt. This is why oil is called a **fossil fuel**. Heat and pressure changed the dead biomass into oil. The sand and silt became rock, so the oil was trapped underground. Drilling into the rock allows the oil to flow up to the surface.

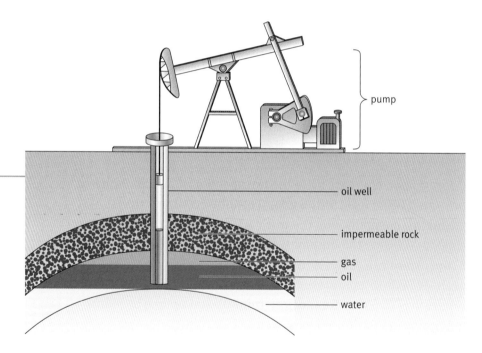

'Nodding donkey' oil pump.

Oil is made of **hydrocarbons** – chemicals made by photosynthesis long ago, containing hydrogen and carbon atoms. So when we burn fossil fuel we are releasing energy from fossil sunlight. The energy was stored as buried **fossil sunlight energy** for millions of years.

When will oil run out?

Oil takes millions of years to form. It is being used far more quickly than it can be replaced.

Some **crude oil** is easy to extract. We have already taken most of this. Oil is rapidly becoming more difficult and expensive to extract. We will stop extracting oil only when other energy sources become cheaper. But alternatives to oil are still some way off.

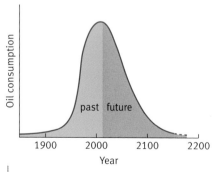

It is not known exactly how long oil will last. Production has passed its peak and may soon start to decline.

Linear systems and oil

The oil economy is a take–make–dump linear system. We take crude oil, refine it to make fuel, and dump carbon dioxide into the atmosphere. We also dump other products made from oil, such as plastics and synthetic fabrics. These products are not reused and they are not biodegradable. Oil is not renewable within our lifetime. Use of oil is not part of a closed-loop system. Waste from oil products accumulates in ecosystems, and is harmful to humans and wildlife.

Carbon dioxide

Carbon dioxide is taken up by plants in photosynthesis. If the concentration of carbon dioxide goes up slightly, plants grow more quickly and bring it down again. This natural **negative feedback** system regulates itself, rather like our body temperature. But, like our body temperature, the system can break down.

As oil is burnt for fuel, carbon dioxide is released. We are now releasing too much carbon dioxide for plants to mop up. Carbon dioxide levels are rising. This is beginning to cause atmospheric temperatures to rise. When carbon dioxide dissolves in water, it produces a weak acid. This makes the seas more acidic, harming marine life such as corals. Corals use carbon dioxide in sea water to make their coral reefs, so removing them adds to the problem.

Climate change threatens ecosystems worldwide. Wildlife may not be able to move or adapt. Melting ice causes sea levels to rise, threatening cities. Climate change does happen naturally. Only 10 000 years ago there was an ice age. But the changes we are causing are more rapid than any recent natural events.

Questions

1 Explain what is meant by a 'fossil fuel'.

2 Oil is made constantly from minute plants and animals, which were built from carbon that was fixed in photosynthesis. When we burn oil, this returns the carbon in these organisms into the atmosphere. Explain why this system cannot be regarded as a closed loop.

The Greenland ice sheet is likely to melt because of climate change, causing sea levels to rise.

Find out about

- **dependence on oil energy for food production**
- **biofuels**
- **models for sustainable living, using recently captured energy from the Sun**

How can the world's people be fed sustainably? To survive without fossil fuels, energy has to come from recent sunlight, not fossil sunlight.

Energy for producing food

Food for today's large human population is produced using energy from oil. Energy is needed for:

- ploughing and planting crops
- making fertilisers and pesticides
- food processing
- transport and distribution of food.

Intensive farming causes pollution, soil erosion, loss of biodiversity, and climate change. It is a take–make–dump, linear system.

Traditional methods

Many parts of the world still use pre-oil farming methods. Fields are cultivated using horses or oxen. The animals' energy comes from biomass (recent sunlight energy) and not from oil. The animals' faeces enrich the soil. Crops are harvested and separated using hand tools. Food is produced and consumed locally. This is a closed-loop sustainable system.

But traditional methods cannot produce enough food for the millions of people living in cities. Traditional methods also leave little time for other activities.

Are biofuels the answer?

Why not harness energy from the Sun by growing oil on farmland? Many plants make oil in their seeds. These can be harvested and turned into biofuel. Biofuel is fuel made from crops and is carbon neutral. The same amount of carbon dioxide fixed in photosynthesis is released when the biofuel is burnt.

Biofuel is now added to petrol. There are targets to increase its use in order to fight climate change. But there are some unwanted effects of growing biofuels:

- biofuels take land needed to produce food
- forests, grasslands, and other wild places are being lost to biofuel crops.

There is no simple pathway to sustainable food production, but biofuels can play a part.

Rice winnowing by hand in India.

It would take the whole of the cultivated land of the USA to grow enough biofuel to meet the country's needs.

Closed-loop systems in industry

There is a drive to change industry from take–make–dump to take–reuse–recycle as in natural ecosystems. Waste from industry can be reused, so that it is a 'technical nutrient', which goes back into the system. Some companies already produce goods sustainably in this way. This is called 'cradle-to-cradle' manufacturing. Goods, such as furniture, computers, and refrigerators are made of reusable materials. Toxic materials are avoided. After their useful life goods are broken down into technical nutrients for re-making.

We don't always need to own all our goods. Leasing cars, bicycles, washing machines, and furniture could make the manufacturer responsible for reusing and remaking. Production–recovery–remanufacture is a closed-loop system, similar to natural systems. Leasing the goods to people helps to control this cycle.

In a sustainable model for living, only biodegradable waste would be returned to the soil. Food made within biological closed-loop systems would feed the population. Energy for manufacturing would come from recent sunlight or other sustainable energy sources, not fossil fuels. Resources would be conserved in technical, closed-loop systems.

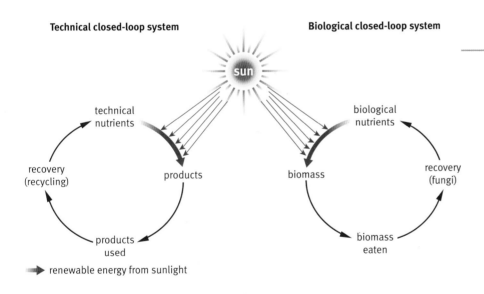

Biological and technical closed-loop systems.

To change to sustainable living, the next generation of scientists, politicians, economists, and businesses will have to apply the model of closed-loop systems in nature to human systems. However, this will need to take into account the needs of local human communities too. This will require research, imagination, determination, and education.

Questions

1. What is biofuel?

2. Explain the term 'carbon neutral' with reference to biofuels.

3. Explain why resources last longer using closed-loop technical systems rather than linear systems.

Topic B7.5: New technologies

The bacteria growing in this tank have been genetically modified to make a protein for use in drug manufacture.

Our scientific understanding of the world changes all the time as scientists make new discoveries and find new solutions to problems. New technologies are based on scientific ideas, but new processes, tools, and machinery take years to develop.

Understanding the science behind the technology that affects your life helps you to evaluate the claims that others make. It also helps you to choose, when faced with an ethical decision.

The Science

To make many of the chemicals we use daily, we grow bacteria and fungi on an industrial scale. We can change the genetic information in microorganisms and plants to make them even more useful to us.

DNA technology helps us to identify individuals, track diseases, and gives evidence that could solve crimes.

Some scientific ideas are regularly in the news. This topic introduces nanotechnology and stem cell technology that could have a great impact in your future.

Ideas about Science

We develop new technologies to improve our lives, but some applications of science can have impacts that we did not predict. Making decisions means balancing benefits against cost. In some areas of science, decisions about what we can do are made by official regulatory groups.

Nothing is completely free of risk. To make a good assessment of risk we need to know what damage might happen and how likely it is to happen.

Find out about

✓ **how microorganisms can be grown to produce useful chemicals**

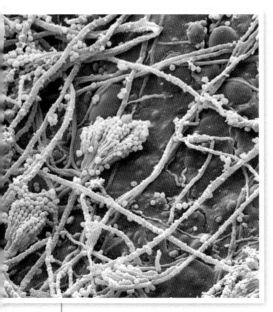

The fungus *Penicillium*. It produces penicillin and secretes it outside its cell.

Early stages of cheese production – churn milk.

There is an amazing variety of microorganisms. Each species produces different **enzymes** and other proteins. Some produce **antibiotics**. Many of these molecules are difficult to make by conventional chemistry. Scientists select species and strains of microorganisms that are able to produce useful chemicals. They then grow the microorganisms in conditions where they will produce chemicals in large quantities. The chemicals are then extracted for use.

Microorganisms are usually grown in batches using huge industrial tanks called **fermenters**. It is difficult to keep the conditions for **fermentation** right for the microorganism. Fast-growing microorganisms use up a lot of oxygen and nutrients. They produce toxic waste products and heat. The conditions inside the fermenter must be carefully monitored and controlled.

Feature	Benefit
Rapid reproduction	Large quantities of products can be made rapidly in a fermenter.
Presence of plasmids	New genes can be introduced into the plasmids in the lab so the bacteria make what we want.
Simple biochemistry	The way that bacteria work is well understood, so the nutrients and growth conditions in a bioreactor can be controlled for optimum production.
Lack of ethical concerns in their culture	There are no animal welfare issues; many processes are similar to age-old brewing and any bacteria are usually removed from the final products.
Ability to make complex molecules	Bacteria can make complex antibiotics, food additives, and hormones that can't be easily synthesised in the lab.

Antibiotics

Antibiotics can treat certain infections. The **fungus** *Penicillium* secretes the antibiotic penicillin. This kills bacteria in its immediate environment. In optimum conditions *Penicillium* can double its mass every six hours. When the fungus grows in a tank of nutrient solution, the antibiotic is secreted into the solution. It is then easy to extract the antibiotic for use as a human medicine.

Harnessing enzymes

Enzymes from microorganisms are very important in food production. They are used to control the flavour, aroma, texture, or rate of production for many food products.

All young mammals make enzymes that help them digest their mother's milk. These enzymes cause milk to form solid lumps so it moves more slowly through the gut. This gives more time for other enzymes to digest the food and for useful molecules to be absorbed.

An extract of enzymes from calves' stomachs is called **rennet**. Rennet is used to make some kinds of cheese. One of the enzymes in rennet is called **chymosin**. Scientists have developed strains of fungus that make chymosin. Some 'vegetarian' cheese is made with fungal chymosin.

Microbial enzymes are added to detergents to make 'bio' laundry liquids and powders. These enzymes digest the fats, carbohydrates, and proteins in the stains on our clothes. 'Bio' detergents often give good results at lower temperatures.

Enzymes to make biofuels

Many scientists are working to produce alternative fuels to fossil fuels. One of these is ethanol, which is an alcohol. We can make ethanol by fermenting the sugars in plant crops such as sugar cane. But crops like this might be needed to feed animals or humans.

Wood is made up of plant cells with cellulose cell walls. Tough fibres called lignin fill the spaces in the cell walls and make the cellulose hard to digest. Scientists have developed a commercial way of making an enzyme, called lignocellulase that breaks down lignin and cellulose. This enzyme can be used to turn woody stalks and leaves into sugars. This means **biofuel** can be made from waste plant material instead of useful crops.

Growing microorganisms for food

Microbial cells contain the same building blocks as cells from other organisms – carbohydrates, fats, and proteins. Some microorganisms produce proteins that are similar to the proteins in fish or soya beans. Microorganisms can be grown on simple nutrients, and they can reproduce rapidly in the right conditions. This means they could be used as food for people or farmed animals.

Quorn is made from a fungus that grows as a cluster of interwoven fungal threads. The threads are extracted from a fermenter, pressed together, and processed to match the taste and texture of meat. Quorn, and some other **single-celled proteins,** have been cleared for human consumption. You can buy Quorn as mince, burgers, and sausages.

Rennet causes the milk to form into solid lumps – the start of cheese.

Quorn is an example of a **single-celled protein** – it is made by fungi. Quorn is nutritionally similar to meat but lower in fat.

Questions

1 Produce a flow chart to explain the main steps in the fermentation of microorganisms to produce antibiotics.

2 Why is it important to control the conditions inside a fermenter?

3 Name three categories of useful product that can be produced by fermentation using microorganisms. Give one example of each.

Key words

✓ enzymes
✓ antibiotics
✓ fermentation
✓ fungus
✓ rennet
✓ chymosin
✓ biofuel
✓ single cell protein (SCP)

Find out about

✓ **genetic modification of bacteria and plants**

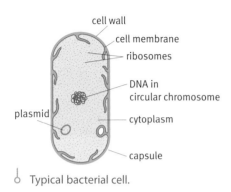

cell wall
cell membrane
ribosomes
DNA in circular chromosome
plasmid
cytoplasm
capsule

Typical bacterial cell.

Plants are sprayed with chemicals to reduce disease and pest damage. The chemicals are expensive to make and can cause pollution.

PRIVATE LAND
GMO TRIAL
KEER OFF

GM varieties of crops could transform farming by cutting the use of chemical sprays.

Bacterial cells make proteins

Bacterial cells are about one-tenth the length of typical animal cells. They have no nucleus, but their DNA forms one large circular chromosome in their cytoplasm. Most bacteria also have small rings of DNA that contain extra genes. These DNA rings are called **plasmids**.

Bacteria produce some proteins that do similar jobs to human proteins, for example, proteins in cell membranes. But bacteria don't make many proteins that are identical to those in human cells. Any cell can make a protein but only if it has the gene that codes for it. Scientists can add a human gene to bacterial cells so that they make a human protein.

Genetic modification

Changing the genes of an organism is called **genetic modification (GM)**. Bacterial cells can be modified by adding genes from other microorganisms, plants, or animals. People do not have the same ethical concerns about modifying bacteria as they do animals or plants.

Many drugs used to treat diseases are proteins. An example of such a drug is **insulin**. Before genetic modification, insulin to treat diabetes was extracted from animals such as pigs. This worked, but pig insulin could produce harmful side effects. GM bacteria can be made that produce insulin identical to human insulin, and that does not cause these side effects.

Genetic modification of plants

Despite producing new varieties of crops by selective breeding over thousands of years, pests, diseases, and weeds still reduce crop yields by about one-third. Genetic modification means developers can add new genes to plants. These genes code for new proteins to give the plant desired properties. Some genetically modified plants are **resistant** to some **herbicides**. Farmers can use these herbicides to kill weeds without harming the crop. However, this may mean that farmers use more herbicides.

Putting new genes into cells

A **vector** is needed to carry the gene into the cell. To modify bacteria, scientists use bacterial plasmids as vectors. Plasmids are easier to manipulate than a bacterial cell's main chromosome. They are small and they move easily in and out of cells.

Not all the cells in a population of bacteria will take in the added plasmid. Scientists put a second gene into the plasmid to make the genetically modified cells easy to select. For example, there is a gene in jellyfish that codes for a green fluorescent protein and several genes make bacteria resistant to particular antibiotics.

To produce human insulin from bacteria, scientists carry out the following steps:

- Isolate the gene for human insulin and make copies of it.
- Make a modified plasmid that contains the human insulin gene and another gene that gives resistance to an antibiotic.
- Add the modified plasmid to a population of bacteria.
- Treat the population with the particular antibiotic.
- The bacteria that survive must contain the plasmid, so they will also make insulin.
- Grow these modified bacteria and harvest the insulin.

Bacteriophages are **viruses** that can infect bacteria. Scientists use them as vectors to carry larger genes into bacterial cells.

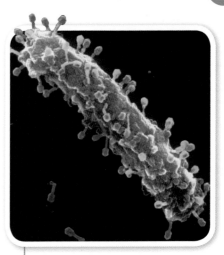

Bacteriophage attacking a bacterium (x15 000).

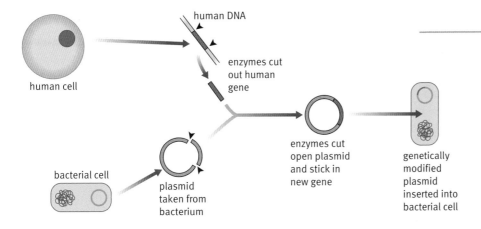

human DNA

enzymes cut out human gene

human cell

bacterial cell

plasmid taken from bacterium

enzymes cut open plasmid and stick in new gene

genetically modified plasmid inserted into bacterial cell

Plasmids are used as vectors in genetic modification of bacteria.

Questions

1 Draw a labelled diagram of a bacterial cell. Highlight the parts of the cell that contain genetic information.

2 Give one example of genetic modification in:
   a bacteria
   b plants.

3 Write a flow chart to show the main stages in the genetic modification of bacteria. Highlight the vector in the process.

4 Which features of bacteria make them ideal for industrial applications such as producing insulin?

5 When genes are inserted into plasmids, marker genes are also added. Explain how the markers are used.

6 Which, if any, ethical issues can you see in genetically modifying bacteria to make useful products?

Key words

- ✓ **plasmid**
- ✓ **genetic modification (GM)**
- ✓ **insulin**
- ✓ **resistant**
- ✓ **herbicide**
- ✓ **vector**
- ✓ **virus**

Philippe Vain.

Philippe Vain is a plant biotechnologist at the John Innes Centre in Norwich. His team, along with researchers at the University of Leeds, has been designing genetically modified crops aimed at helping many of the world's poorest farmers. 'Our goal is to improve the pest resistance of key crops – rice, bananas, and potatoes – for developing countries', says Philippe. 'In 10 years there will still be more than half a billion people in the world without a reliable source of food. It is a much better strategy to give these people the means of food production instead of supplying food aid all the time.'

Nematode worms reduce crop yields

The target of Philippe's research is nematodes – microscopic worms that live in the soil. These worms attack the roots of crops, taking nutrients from the plant and laying their eggs inside the tissues. 'If you have a small infestation, you're going to get a reduced yield, a large infestation, and you'll lose most of the crop.' For a poor farmer this can be a matter of life and death.

Farmers could kill the worms by spraying the crops using chemicals called nematicides, but these are expensive and highly toxic to humans and the environment. Instead, it was decided to develop a crop that was resistant to the pests.

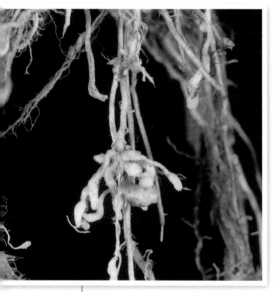

Philippe works on a species of root-knot nematode worm called *Meloidogyne*.

Adding an extra gene

The plants already have genes for natural substances called cystatins. The cystatin genes are active in certain parts of the plant, such as their seeds. Cystatins affect insect digestion, so insects cannot eat parts of the plant that contain them. Cystatins have no effect on humans. In fact we eat them all the time, in seeds from crops such as rice and maize.

When a particular gene is active in a cell we say that it is being expressed. This means that the protein for which it codes is being made in the cells. Philippe's team added another copy of the cystatin gene to the plants, which is expressed in the root cells.

Agrobacterium species cause cancer in plants. This plant is infected with *Agrobacterium tumifaciens*, which causes crown-gall disease.

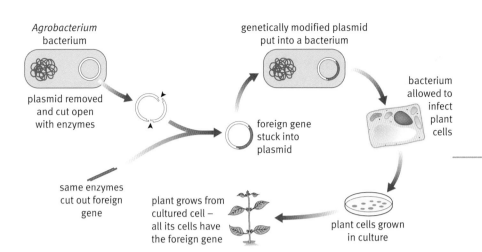

Agrobacterium bacterium

plasmid removed and cut open with enzymes

same enzymes cut out foreign gene

foreign gene stuck into plasmid

genetically modified plasmid put into a bacterium

bacterium allowed to infect plant cells

plant cells grown in culture

plant grows from cultured cell – all its cells have the foreign gene

Agrobacterium is used to transfer a gene into a plant's DNA. The bacterium's plasmid acts as the vector in this example of genetic modification.

The cystatin gene makes the roots indigestible to the nematodes. The researchers used the bacteria *Agrobacterium* as a vector to carry the extra gene into the plant's genetic material.

The only difference between the final genetically modified plant and the original is an extra copy of the cystatin gene. Nevertheless, by law, any genetically modified plant has to go through extensive testing and safety trials before it can be released into the environment.

The resulting plants show a high level of resistance to the nematode and are ready to be offered to farmers as part of a government aid project. Phillipe Vain: 'You want to make a contribution. It's very rare to have a crop improvement strategy that really works, so it's very exciting to see the outcome.'

'For us, the best result will be people trying the crop and it making a difference to their lives.'

Much of Philippe's work is on banana plants and rice.

Question

1 What is the vector for the foreign gene in this case?

Find out about

- ✔ how the use of GM crops in the UK is controlled
- ✔ some of the risks and benefits of GM crops

Genetically modified soya.

Making decisions about GM

To introduce a genetically modified organism (GMO) into the UK environment an application must be made to a government department. Each application is looked at by an outside agency. This agency advises the government on every GMO application. At the beginning of 2010, there had been some field trials of GM crops in the UK, but none were grown commercially.

In the USA there is also a series of committees that manage what can be done with GM crops. GM soya, cotton, corn, and canola (a variety of oilseed rape) are grown commercially there. Genetic modifications to crops have improved their resistance to pests and diseases, to reduce spraying and crop losses. Some crops have been genetically modified so that they are resistant to herbicides. Herbicides kill weeds, so the crop has less competition and grows better. When the GM crop is sprayed with the recommended herbicide the only plants left alive are those with the matching herbicide-resistance gene.

Why are some people concerned about GMOs?

GMOs are living things. They reproduce between themselves and interbreed with non-GMOs. It is impossible to predict how GMOs will interact with other species. The potential risks and benefits of introducing GMOs into an ecosystem need to be balanced.

This table outlines some of the arguments about GMOs.

Arguments against introducing herbicide resistance gene into a plant	Counter arguments
Added genes could make 'safe' plants produce toxins or allergens.	Food-safety organisations can check for these.
Marker genes for antibiotic resistance could be taken up by disease organisms.	The antibiotics are not used in medicine, so it wouldn't matter.
Pesticides could 'leak' out of the roots of GM plants and damage insects or microorganisms that they were not designed to kill.	Insect-resistant plants reduce pesticide application so they have an overall benefit to the environment.
GM crops may cause changes to ecosystems that cannot be reversed.	New crops produced by selective breeding have not caused huge changes to ecosystems so far.
It will cost farmers more to buy seeds of GM crops, so food costs will increase.	Farmers may benefit from healthier crops and lower costs of production.
Multinationals will increase their domination of world markets.	Some GM technology has already been shared with developing nations.
Many consumers in EU countries refuse to buy GM products so farmers may lose markets.	Consumers in most countries would buy GM crops.
Poor farmers will not be able to afford the GM seeds.	Gene technology could develop more nutritious, higher-yielding or drought-resistant crops that could benefit developing countries.

Evaluating evidence for and against GMOs

The arguments about GMOs are complicated. This is partly because it is a new technology and people cannot be certain about predicting the outcomes. Also, the technology involves living things in environments where many factors can affect the results.

Most of the information available about GMOs is produced by people who feel very strongly about the issue. They might focus on the benefits to promote the development of GMOs or the potential hazards to discourage the use of GMOs. When you find information about genetic modification, it is important to know who has produced it. This helps you to decide how much to trust their judgment.

On big issues like GM crops, the government makes decisions for us. They make **regulations** and laws allowing or forbidding new developments. Companies with an economic interest in making new products try to influence decision makers. Members of the public and action groups also have an influence. Outside agencies try to make sure they have independent, scientifically reliable information, so that the government can make an informed decision.

Genetically modified cotton.

Questions

1 A herbicide-resistance gene is added to a crop plant. What advantages could there be for the:
   a seed/herbicide seller?
   b farmer in the developed world?
   c farmer in the developing world?
   d environment?
   e consumer?

2 What possible disadvantages or risks could there be for each group in question 1?

3 How could potential risks of new technologies be reduced?

4 What evidence would convince *you* that the risks of GM organisms are worth taking, in order to have the benefits of this new technology?

Key word
✓ **regulations**

Find out about

✓ how genetic fingerprinting works – the technique from which genetic profiling has been developed

Sir Alec Jeffreys in the 1980s.

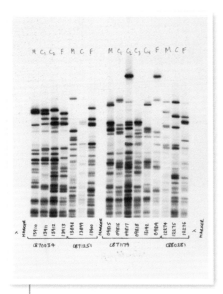

Each black line shows where pieces of DNA contain minisatellite sequences. They were separated out by gel electrophoresis.

'My life changed on Monday morning at 09:05 am 10 September 1984. In science it is unusual to have such a "Eureka moment". We were getting extraordinarily variable patterns of DNA, including the DNA of our technician and her mother and father. My first reaction to the results was "this is too complicated"; then the penny dropped and I realised we had genetic fingerprinting.'

Sir Alec Jeffreys is describing the discovery that made him one of the most famous scientists in the world. Firstly, his research group broke cells in the tissue sample and purified the DNA. They cut the DNA into fragments, using enzymes. Sir Alec Jeffreys and his research group developed a short length of DNA called a **probe**. Probes attach to complementary sections of DNA. The probe they used stuck to DNA in minisatellite sequences. These sequences repeat thousands of times throughout a person's DNA. The number and pattern of repeats varies between individuals.

```
CGCCTCGGCCTCCCAGAGTGCTGAGATTACAGGCGTGAACCACCATGCCTAGCCGTTAGCTCCCA
CTTATGAGTGAGAACAGGTGATGTTTGGTTTTCCATTCCTGAGTTACTTTACCCAGAATTGTTGT
CTCCAATCTCATCGAGGTCTCTGCGAATGCCAGTAATTCATTCCTTTTTATGGCTAAGTAGTATT
CCATCGTATATATACATATACATATATATGTATACACACATATACATATATATGTATACACACAT
ATACATATATATGTATACACACATATACATATATATGTATACACACATATACATATATATGTATA
CACACATATACATATATATGTATACACACATATACATATATATGTATACACACATATACATATAT
ATGTATACACACATATACATATATATGTATACACACATATACATATATATGTATACACACATATA
CATATATATGTATACACACATATACATATATATGTATACACACATATACATATATATGTATACAC
ACATATACATATATATGTATACACACATATACATATATATGTATACACACATATACATATATATG
TATACACACATATACATATATATGTATACACACATATACATATATAGTATACACACATATACATA
TATATGTATACACACATATACATATATATGTATACACACATATACATATATATGTATACACACAT
ATACATATATATGTATACACACATATACATATATATGTATACACACATATACATATATACATATA
TACACAAACACATGCACCGCACTTTTTTTTTTTTTTTTTTTTGAGATGGGAGTCTCACTCTAT
CACCAGGCTGGAGTGCAGTGGTGTGATCTTGGCTCACTGCAACCTCTGTCTCCTGGGTTCAAGCT
ATTCTCCTGCTTCAGCCTCCTGAGTAGCTGGGATTACAGGTGCTCACCACCATGCCCAGCTAATT
TTTGTATTTTTACCATGTTGGCCAGGATGATCTCCATCTCCTGACCTCCTGATCCTCCTGCCTTG
```

Part of the DNA sequence from chromosome 21, showing repeating minisatellite DNA sequences.

Fragments of DNA were placed on a gel with an electrical field across it. This is called **gel electrophoresis**. Fragments move through the gel at different rates, according to their size. Fragments of the same size move together, and form a band. Radioactive probes latched on to minisatellite sequences and showed the bands as dark lines on X-ray film. Each person's DNA gives a different pattern of bands. Closely related individuals have some bands in common.

Developing genetic profiling

The first DNA fingerprints were difficult to interpret and needed quite large samples of high-quality DNA for the technique to work. In some situations, for example, at crime scenes, there are only small samples of DNA. These may be in a few drops of blood or a hair follicle. If the sample is old, it may have decomposed.

Since 1985, techniques used to analyse DNA from different sources, have developed. **DNA profiling** is still done using electrophoresis, but now with gels in narrow capillary tubes. Profiling highlights around 20 different repeating sequences in a DNA sample. The differences between samples are shown as differences in the patterns of multiple repeats in these regions.

Bands of DNA in gels are made visible with marker chemicals. At first, radioactive markers were used that could be detected with X-ray film. Now, safer fluorescent markers are used, which glow (fluoresce) when stimulated by light. Computers interpret the patterns of bands in gels and produce a printout. Scientists no longer study and measure the gels directly.

Modern DNA profiling can be carried out with smaller samples of DNA than earlier methods. Even a very small spot of blood will contain DNA in the white blood cells. The DNA from samples can be copied many millions of times using a technique called a polymerase chain reaction (PCR).

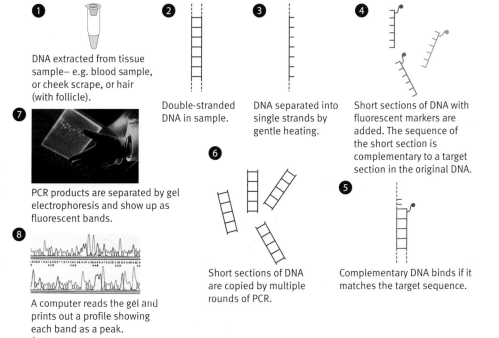

1 DNA extracted from tissue sample– e.g. blood sample, or cheek scrape, or hair (with follicle).

2 Double-stranded DNA in sample.

3 DNA separated into single strands by gentle heating.

4 Short sections of DNA with fluorescent markers are added. The sequence of the short section is complementary to a target section in the original DNA.

5 Complementary DNA binds if it matches the target sequence.

6 Short sections of DNA are copied by multiple rounds of PCR.

7 PCR products are separated by gel electrophoresis and show up as fluorescent bands.

8 A computer reads the gel and prints out a profile showing each band as a peak.

Flow diagram of genetic profiling methods.

Key words

- ✔ **probe**
- ✔ **gel electrophoresis**
- ✔ **DNA profiling**

Question

1 What are the main advantages of genetic profiling over the early DNA fingerprinting techniques?

Find out about

- ✓ **how genetic testing makes use of DNA technology**

What does a genetic test do?

Everybody's DNA is different. Genetic tests using DNA profiling identify sequences at particular places in a DNA sample. Genetic tests are used to identify an individual, for example, to match someone to the DNA found at a crime scene.

For each of us, any part of our DNA has come from either our mother or our father. A genetic test can show family relationships between people. For example, it can show that a particular man is the father of a child. Genetic testing is also used to identify human remains and can identify criminals.

DNA profiling was first used to prove a person's innocence. In 1985 a man confessed to the brutal murder of a young girl. However, DNA fingerprinting proved him innocent. The man responsible was caught when DNA samples were taken from the local male population and analysed by profiling.

Modern laboratory procedures make it much easier to find out the exact sequence of bases in a section of DNA. It is possible to find out if a gene in a DNA sample is a variant of the gene (allele) that is linked to the genetic disorder. This makes it possible to identify affected individuals or carriers of a disorder or disease with a genetic component.

DNA profiling works on DNA from other species too. It is used to study populations of endangered animals to find out how closely related the individuals are. This information is used to plan breeding programmes in zoos. It can also match individual animals to DNA from crime scenes or to identify ownership.

Developing DNA analysis
Probes for DNA analysis

Genetic testing using DNA technology can identify genes associated with diseases in a person's DNA. Some genetic tests use specially made pieces of DNA called **gene probes**. A gene probe is a short piece of single-stranded DNA, just like the probe used by Alec Jeffreys, but that has a sequence of bases complementary to the gene being tested for. As a result, the probe will bind to the DNA in the sample if the gene is present.

Archeologists at work. Genetic testing can provide information about people who died a long time ago.

The impact of DNA profiling

DNA profiling does not directly solve crimes. Just because a person's DNA is present at a crime scene does not mean that they committed the crime.

Government, lawyers, and civil-liberties action groups have been discussing the ethical position of the UK DNA database. In England and Wales in 2010, the law permits the police to take DNA samples from anyone arrested. Those samples and their DNA profiles can be kept in the database even if the person is not convicted of any crime. Some people argue that this means criminals can be quickly identified and innocent people eliminated. Others argue that keeping DNA, or DNA profiles, is not a good idea as the information could have other uses in future.

Individuals may find it interesting or useful to learn about their genetic profile and their likelihood of contracting particular diseases. However, it could also be possible for employers and insurance companies to use genetic test information to make decisions affecting people's lives.

How are genes copied?

PCR (the polymerase chain reaction) is now central to DNA analysis. Using PCR scientists can make millions of copies of a selected piece of DNA in a few minutes.

PCR is used:
- in forensic science when the amount of DNA found at a crime scene is very small
- to copy a gene for genetic modification
- to make gene probes
- to make many copies of a region of interest so that it can be studied further – for example, to look for changes related to disease.

Forensic scientists can identify individuals from tiny samples of hair, sweat, or blood at crime scenes.

Questions

1 List four uses of genetic tests.

2 Describe the structure of a DNA probe.

3 How does a DNA probe help you to carry out a genetic test?

4 How do scientists find out which DNA sequence has a gene probe stuck to it?

5 Suggest why blood is a good source for a DNA sample.

Key word
✓ gene probe

What is 'nanotechnology'?

Nanotechnology is a technology that makes use of very tiny particles. The particles used in nanotechnology are as small as 100 **nanometres** in at least one dimension. They are individual particles, not joined together into larger structures. A nanometre is extremely small. There are a million in a millimetre. A cell membrane, for example, is 6–10 nanometres thick.

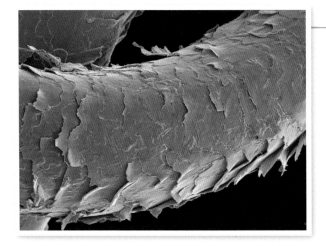

A human hair is one hundred thousand nanometres in diameter.

Some sunscreens have nanoparticles of zinc or titanium oxides.

Nanotechnology meets biology

A wide range of uses for nanotechnology are being developed by biologists.

People have known for centuries that silver can make food last longer. More recently, manufacturers have developed plastic with tiny particles of silver embedded in it. This can be used for making food storage boxes and film wraps for food. The silver particles reduce contamination of the food by microorganisms.

Several companies have developed plastic food wrapping films that change colour when:
- antibodies in the film react with bacteria in the food
- nanoparticles in the film react to changes in the amount of oxygen in the packet, which is a sign that the wrapping could be damaged
- fruit ripens and releases gases that react with nanoparticles in the film.

How does nanotechnology work?

When materials are made with smaller and smaller particles their properties often change. For example, they may conduct electricity better, allow light through, or change from solid to liquid. When you divide a particle, you increase the exposed surface without increasing its volume. Some of the properties of nanoparticles may be the result of this increase in surface area.

Plastic containing nanoparticles of silver has antibacterial properties.

Nanosilver

Nanometre-sized particles of silver can be absorbed into animal cells. Once inside the cells, nanosilver particles react with cell contents and release silver ions. Silver ions are similar in size to sodium ions and disrupt normal cell activities. Solutions of silver salts contain silver ions and can be used directly for their antibacterial effect. But ions in a silver salt solution react with other molecules before they can be absorbed. This means you would need to use higher concentrations of silver in a silver salt solution to get the same antibacterial effect as a nanosilver product.

Questions

1 What is nanotechnology?

2 What are some of the potential risks of nanotechnology?

3 What do you think are the chances of nanotechnology damaging the environment? How severe would the consequences be if unexpected problems happened?

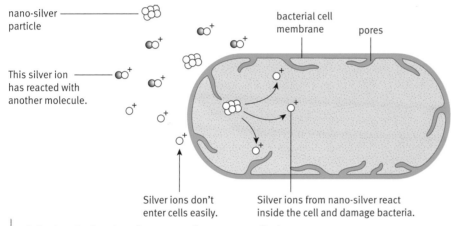

nano-silver particle

This silver ion has reacted with another molecule.

bacterial cell membrane

pores

Silver ions don't enter cells easily.

Silver ions from nano-silver react inside the cell and damage bacteria.

Cells absorb silver ions from nanosilver more easily than silver ions in silver salt solutions.

Key words

✓ **nanotechnology**
✓ **nanometer**
✓ **risk**

Risks and benefits

Technological developments are often made with the goal of improving lives. However, any new technology may bring with it an unpredicted hazard or **risk**. Nanometre-sized particles may behave in unexpected ways or be toxic to humans. For example, there have been some concerns that silver in food packaging could leak out into landfill or into waterways and cause environmental damage.

Current safety measurements and risk assessments are based on chemicals in their more usual form. We may need to revise our ideas of what is safe and what is not safe when working with nanometre-sized particles of any chemical. As with any new technology, it is important that systems are in place to test its safety before it is widely used.

Nanosilver film ends up in landfill sites like this one, but where does the nanosilver go?

What are stem cells?

Most of the cells in our bodies are **differentiated**. This means they have become specialised to do a particular job in our body. For example, nerve cells (neurons) are long strands connecting two points in the body. They produce an electrical charge that carries a signal from one place to another.

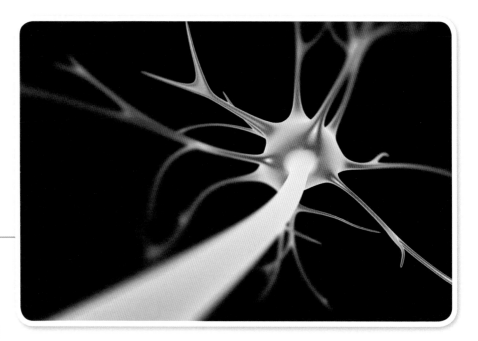

Neurons carry nerve impulses through your nerve network.

Differentiated cells cannot make copies of themselves to repair or replace damaged tissues. **Stem cells** in specialised tissues have the potential to develop into a range of different cells. For example, skin stem cells can multiply and differentiate to repair damaged skin **tissue**.

Medical treatments with stem cells

For over 30 years, blood disorders, such as **leukaemia,** have been treated with bone marrow transplants. Bone marrow contains stem cells that divide and differentiate to make eight different kinds of blood cell. Leukaemia is a kind of cancer where the body makes too many white blood cells. Treatment involves killing the patient's own bone marrow cells with radiation. New bone marrow introduced from a donor can make healthy blood.

More recently scientists have carried out **tissue culture** in the laboratory. Stem cells are grown in a special growth solution containing proteins and sugars to stimulate growth. Skin tissue culture produces a thin layer of skin cells that can be used as a skin graft to treat burned skin.

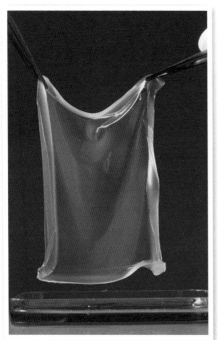

Scientists grew this skin from skin stem cells in sterile conditions. Doctors use it for skin grafts. A skin graft from a person's own stem cells reduces the risk of rejection.

Potential of stem cells

Researchers are working to see if stem cells can be used to repair damaged spinal tissue. Potentially this could restore movement to people paralysed by spinal cord injuries. Researchers inject stem cells into the damaged area in some of their patients or experimental animals and compare their recovery with patients who have different treatments.

It may be possible in the future to repair brain damage, such as that caused by Parkinson's disease or Alzheimer's. Stem cells could be used to treat some kinds of diabetes, particularly in young people, if the cells can develop in the pancreas to produce insulin.

It is difficult to develop a complete **organ** in the laboratory, as organs are such complicated structures. In our bodies, each organ works because it is connected to our circulatory system and includes a network of capillaries. Growing complete organs from stem cells will need a lot of research.

Where do stem cells come from?

All our bodies contain some stem cells, for example, in our skin, blood, and bone marrow. For some conditions, the patient might be the best donor. If a disease or condition is caused by a faulty gene in the patient's cells, another closely matching donor is needed.

Umbilical cord blood is one source of stem cells. Each baby born could provide some stem cells for research and future treatments. The many embryos left over from fertility treatments are another potential source of stem cells. There are regulations controlling how these embryos can be used. Only very early embryos, made up of a few cells, can be used for research. With special treatments in cell culture, ordinary differentiated body cells can sometimes be made to behave like stem cells. If stem cells can be reliably made in this way, it would solve many of the ethical problems surrounding the use of embryos.

Questions

1 What is the important difference between stem cells and differentiated cells?

2 Are stem cells a completely new technology?

3 What new treatments could stem cells be used for?

4 What do you think about using bone marrow, umbilical cord blood, or embryos as sources of stem cells?

Key words

- ✓ **differentiated**
- ✓ **stem cells**
- ✓ **tissue**
- ✓ **leukaemia**
- ✓ **tissue culture**
- ✓ **organ**

Very early human embryos, such as this, can provide embryonic stem cells.

Find out about

- ✓ how hearts can fail
- ✓ how we design and engineer replacements for faulty heart valves
- ✓ how pacemakers are used to restore normal heart function

How does the heart work?

Look back at page 214 in this book to remind you of how the heart works. **Heart valves** make sure the blood flows in one direction only. An area of the heart called the **pacemaker** controls the muscle contractions, using electrical signals. This makes sure the heart muscle contracts in the right sequence and pace. It keeps your heart beating at a slow and steady pace when you are resting and at a higher rate when you exercise.

What goes wrong with our hearts?

Blood vessels run into the heart muscle and provide the muscle with blood containing oxygen and food. These blood vessels can get blocked, for example, by fatty deposits on their lining. When this happens, parts of the heart muscle receive no blood and the muscle tissue quickly dies. This is what happens when you have a heart attack.

Other problems can be caused by the valves in our hearts not working properly. The tissue making up the valve flaps may get stiff or torn with wear, so the valve won't work properly.

Sometimes problems with the pacemaker result in an irregular heartbeat.

Valves in your heart keep the blood flowing in the right direction and electrical signals from the pacemaker keep your heart beat regular.

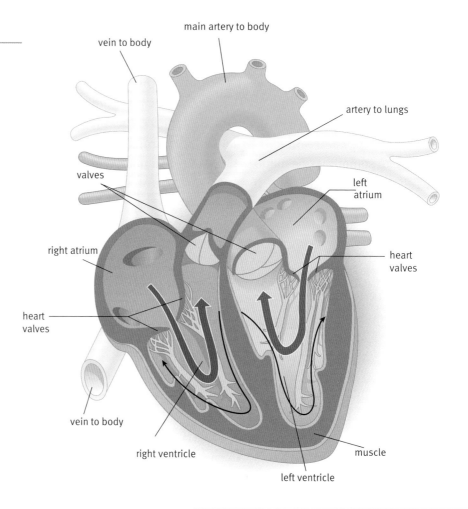

Replacing valves

To replace a heart valve, a surgeon connects the patient's blood supply to a heart–lung machine. Then the surgeon stops the heart, cuts it open, and replaces the damaged valve.

Replacement valves can come from a human or animal donor. The main disadvantage of a tissue transplant is that the immune system may cause **rejection** of the transplant. This happens because our immune system sees the new tissue as foreign, and attacks it as if it were an invading bacteria or virus. Another option is to use an artificial valve that has been engineered to do the job. Disadvantages of these products of **biomedical engineering** are that they cause damage to blood cells and make regular clicking noises as the valve closes.

The metals and plastics used to make replacement body parts such as heart valves have to be resistant to wear and tear. They must also be made from materials that do not corrode in the body, and do not stimulate the body's rejection systems. This avoids patients having regular operations to replace worn out valves. Developing new materials to replace body parts is an important application of chemistry.

Restoring rhythm

If the heart loses its natural rhythm, it might just feel strange or it might make any exercise difficult. An artificial pacemaker can monitor your heart rate and stimulate the muscle to contract in a regular rhythm.

Some scientists have used stem cells to develop muscle tissue in the laboratory. These muscle fibres contract and relax regularly, like a beating heart. If the cells can be encouraged to develop in the same way inside a heart, it might be possible to repair the damage caused by a heart attack. Clusters of these new cells could be used as a natural pacemaker.

Key words

- ✔ **biomedical engineering**
- ✔ **heart valves**
- ✔ **pacemaker**
- ✔ **reject ion**

Questions

1 What is the job of the heart?

2 What do the valves within the heart do?

3 Why might heart valves need to be replaced?

4 What are the main problems following heart-valve replacement surgery?

5 Why is a pacemaker important in your heart's function?

6 If you ever need surgery, would you prefer an artificial valve or a donated valve from another human or an animal? Give your reasons.

These photographs show examples of replacement heart valves.

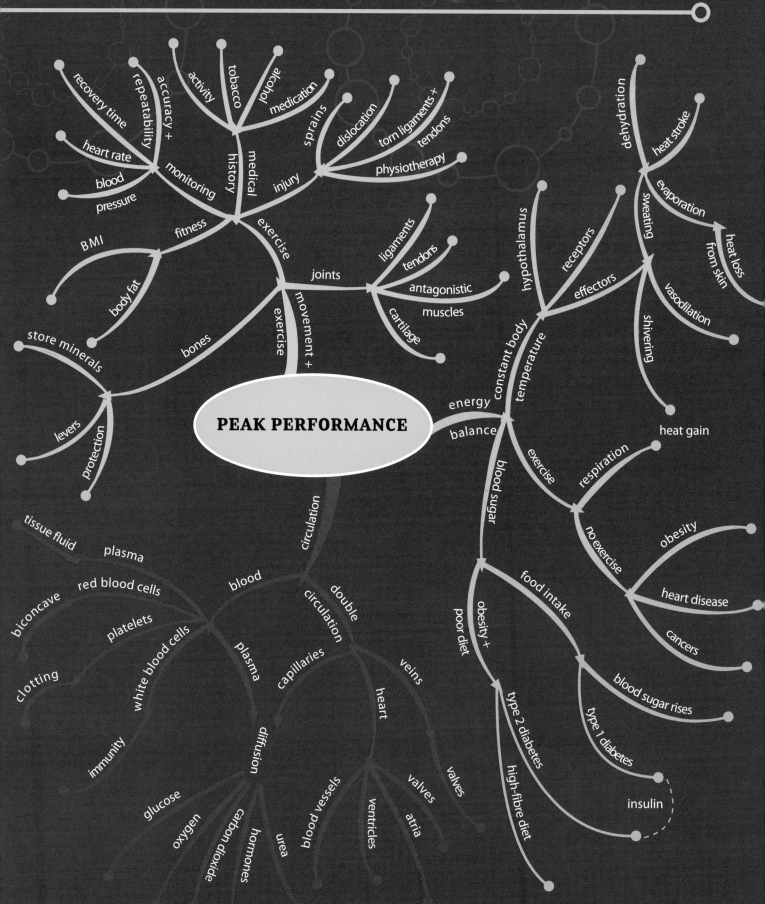

PEAK PERFORMANCE

recovery time
repeatability
accuracy +
activity
tobacco
alcohol
medication
monitoring
heart rate
blood pressure
medical
history
sprains
injury
dislocation
torn ligaments +
tendons
physiotherapy
BMI
fitness
body fat
exercise
ligaments
tendons
joints
antagonistic muscles
cartilage
store minerals
bones
movement +
exercise
levers
protection

dehydration
heat stroke
evaporation
sweating
heat loss from skin
hypothalamus
receptors
effectors
vasodilation
shivering
heat gain
constant body temperature
energy balance
exercise
respiration
blood sugar
no exercise
obesity
heart disease
cancers
food intake
blood sugar rises
obesity + poor diet
type 2 diabetes
type 1 diabetes
high-fibre diet
insulin

tissue fluid
plasma
red blood cells
biconcave
platelets
clotting
white blood cells
immunity
blood
double circulation
circulation
plasma
capillaries
diffusion
glucose
oxygen
carbon dioxide
hormones
urea
blood vessels
veins
heart
ventricles
valves
atria
valves

LEARNING FROM NATURE

soil erosion
silting
Easter Island
deforestation
fewer tigers
eutrophication
biodiversity loss
fertilisers
desertification
climate change
farming
job loss
oil reserves
running out
fossil fuels
timber
fishing
overfishing
collapse of fish stocks
linear
take–make–dump
non-biodegradable
sewage
pollution
bioaccumulation
raw materials
non-renewable

timber
clean air
ecosystem services
water
fish + game
soil
closed loop
ecosystems
sustainable
plants
lifestyle
soil conservation
dead organic matter
animals
rural societies:
Maasai
sustainably managed
post-oil world
well-managed
fish stocks
renewable energy
fungi + bacteria
faeces
dead animals
forests
biofuels
stable human population
new sustainable lifestyles
extensive farming
nutrients
carbon dioxide
cradle-to-cradle
manufacture
no waste

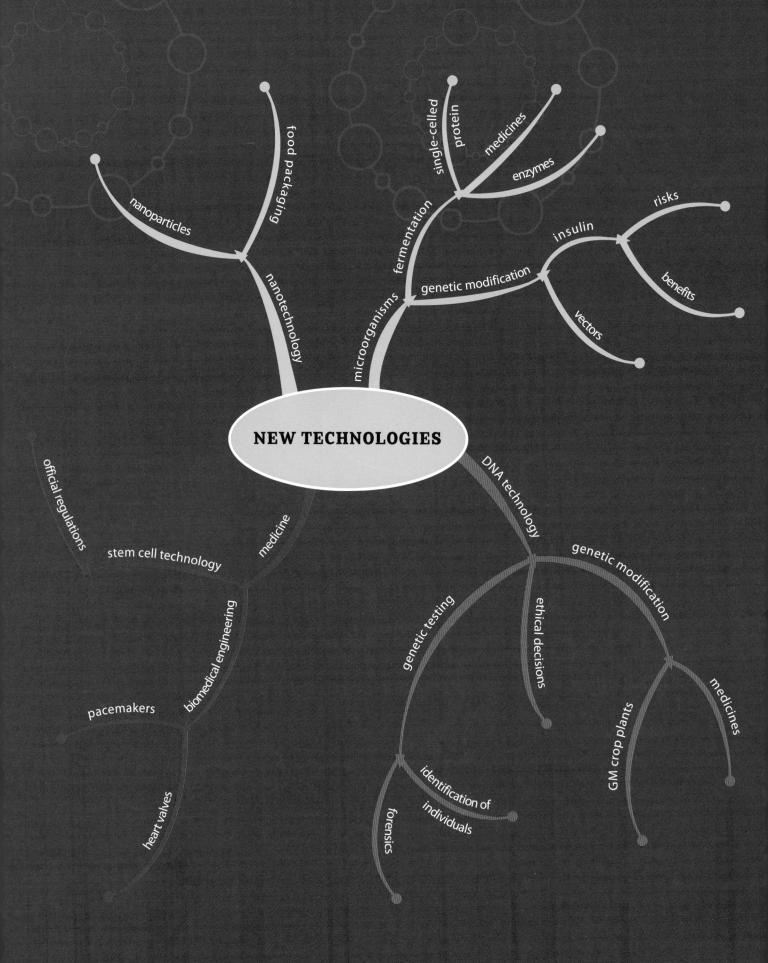

NEW TECHNOLOGIES

nanotechnology
- food packaging
- nanoparticles

microorganisms
- fermentation
  - single-celled
  - protein
  - medicines
  - enzymes
- genetic modification
  - insulin
  - risks
  - benefits
  - vectors

medicine
- stem cell technology
  - official regulations
- biomedical engineering
  - pacemakers
  - heart valves

DNA technology
- genetic testing
  - ethical decisions
  - forensics
  - identification of individuals
- genetic modification
  - GM crop plants
  - medicines

Science Explanations

The human body is capable of remarkable things – climbing mountains, swimming and diving, running, and weight lifting – as well as tasks needing dexterity such as crafts and typing. To get the best from our bodies and to avoid injury it is helpful to know how the body works.

You should know:

- that vertebrates have an internal skeleton, supporting the body and allowing movement
- that muscles can pull but not push, so are arranged in opposing pairs to move joints
- the properties of tendons, ligaments, cartilage, and synovial fluid and how they are arranged in a joint
- why a person's medical history should be considered before an exercise plan is set up
- how to interpret data about heart rate, blood pressure, and recovery period during and after exercise
- how to calculate body mass index (BMI) as a measure of fitness
- that recorded changes in fitness may be affected by monitoring errors and need to be repeated for reliability
- the types, symptoms, and treatments of injuries that can be caused by excessive exercise, and the role of a physiotherapist for sports injuries
- what is meant by a double circulatory system
- the components of blood and their role in transport, fighting infection, and blood clotting
- how red blood cells are adapted to their function of carrying oxygen
- the names and functions of the chambers and main blood vessels of the heart
- the function of valves in the heart and veins
- how materials are exchanged between capillaries and cells
- how humans maintain a constant body temperature
- that foods with high sugar content can lead to a rapid rise in blood sugar level but a high-fibre and high-starch diet causes less fluctuation
- about the differences between type 1 and type 2 diabetes in terms of their causes and treatment, including the role of insulin, diet, and exercise
- how to interpret data on the risks to health caused by an unhealthy lifestyle.

Learning from nature

We are running out of energy and food resources and damaging the planet's ability to sustain life. By studying natural ecosystems we can learn to live in a more sustainable way and leave a healthy environment for future generations.

You should know:
- how ecosystems behave as closed loops
- that ecosystems create no waste because the output from organisms is used as input for other organisms
- some examples of the outputs and inputs within ecosystems
- how fungi and other microorganisms use enzymes to feed on dead organic matter and release nutrients
- how to interpret data and diagrams showing the movement of nitrogen, carbon, and oxygen through ecosystems
- that stable ecosystems, such as rainforests, recycle most materials, with few gains or losses
- why flowers, pollen, seeds, and other reproductive structures are produced in large quantities
- the importance of ecosystem services provided by natural systems
- that humans rely on natural ecosystems for life support in the form of food, clean water, clean air, fish, game, soil, and pollination services
- the consequences of linear, unsustainable systems used in farming, timber production, fishing, and manufacturing industries
- how non-biodegradable waste can accumulate in food chains (bioaccumulation), harming wildlife and possibly human health
- that humans alter the balance of natural ecosystems by changing inputs and outputs, leading to problems such as eutrophication
- why oil production and use is not within a closed-loop system
- solutions to restore the balance in damaged ecosystems caused by fishing, agriculture, and timber production
- how conservation helps to preserve ecosystem services but can be against the short-term interests of local communities
- how human activity can be sustainable if it can harvest enough energy from current sunlight.

New technologies

To evaluate and make sense of new technologies, we need to understand the background science involved. Even with a good understanding of science, some decisions about the applications of new technologies are made for ethical reasons or because of the balance of risks and benefits.

You should know:
- how we manipulate bacteria and fungi to produce useful chemicals
- the process and uses of genetic modification of microorganisms and plants
- how gene probes are used to analyse DNA
- that detailed analysis of DNA allows us to track diseases and to identify individual people or animals
- about nanotechnology in the food industry
- about the ways in which stem cell technology could be used to treat disease
- some biomedical engineering solutions to problems in the heart.

Ideas about Science

Achieving 'peak performance' is not easy. We can try different diets and exercise regimes but we need to know how well they are working. People vary greatly in their health and physical makeup and scientists must take this into account when measuring fitness. We need to know how diet and lifestyle affect our health and fitness. This means finding a correlation between particular foods or exercise regimes and a person's physical condition.

You should be able to:
- explain why measurements of fitness vary, both within an individual and between people
- judge the reliability of a set of data based on the repeatability of the measurements
- interpret where the true value lies in a set of measurements, based on the mean
- suggest whether there is a real difference between two sets of measurements
- take account of outliers and explain reasons for including or discarding these
- understand that correlation means *either* that one factor increases *or* decreases the chance of the other factor occurring, or that the value of one factor increases/decreases in step with the value of the other factor
- recognise and interpret correlations in data
- discuss why a correlation does not always indicate cause.

Science has enabled us to grow more food and harvest more fish and timber. These have enhanced the quality of life for many people, but have also had harmful effects on the environment. We must find alternative, sustainable methods before too much damage is done. Decisions about the exploitation of wildlife should be guided by ethics and regulation.

You should be able to:
- give examples of non-sustainable farming methods, explaining their benefits and disadvantages
- explain why different groups of people might reach different decisions about using these technologies
- identify examples of unintended impacts of human activity and suggest sustainable alternatives
- interpret data to assess the sustainability of different methods of resource use and manufacturing
- discuss the regulation of wildlife, fish, and timber use, and its effectiveness
- distinguish between questions that can be answered using scientific methods and those that can not
- summarise different views about ethical issues
- identify and develop arguments based on different ethical standpoints.

Any new application of science brings with it some benefits and also some risks. Scientific work is officially regulated, and the regulators use their scientific knowledge to make assessments and decisions when the outcomes are uncertain. Some decisions are made using ethical arguments rather than scientific ones.

You should be able to:
- identify risks that might be involved in using new technologies and suggest ways to reduce them
- show awareness of, and discuss, how science and the applications of science are officially regulated
- identify the ethical issues involved in suggested new technologies
- identify who might be affected by a suggested new technology, and discuss the main benefits and disadvantages of that technology.

Questions

1 Muscles work in pairs to move a joint.
  a Give the name for this muscle arrangement.
  b State which muscle contracts and which relaxes to bend the elbow joint shown below.

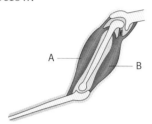

2 **a** Match functions i, ii, and iii to the blood components in the list below.
  Functions:
    i transport oxygen
    ii fight infection
    iii help the blood to clot when injured
  Blood components:
  white blood cells
  red blood cells
  platelets
  b Explain why white blood cells but not red blood cells can be used as a source of DNA.

3 **a** Explain why the body tends to get hotter during exercise.
  b Name the part of the brain that controls body temperature.
  c Explain two ways in which the body can lose heat when the body temperature rises.

4 A person's body mass index (BMI) is 30. Their height is 1.8 m.
  a How many kilograms should the person lose to bring their BMI down to 24?

b Suggest three reasons why it would not be a good idea for the person to immediately join their local jogging group.

5 Which of the following involve linear systems and which involve closed-loop systems?
  a decaying leaves in a forest
  b disposal of domestic waste
  c traditional Maasai village
  d burning fossil fuels
  e commercial fishing
  f food chains in a natural lake

6 Briefly explain why starch is biodegradable and polythene is not.

7 Photosynthesis uses carbon dioxide and respiration releases it.
  a Explain why the concentration of carbon dioxide in the air in a mature forest remains roughly constant.
  b Explain why carbon dioxide in the atmosphere has increased over the past 100 years.

8 A scientist measured the levels of mercury in fish and birds living in a polluted bay. Small herbivorous fish had a concentration of 1 unit in their livers. Larger predatory fish had a concentration of 100 units and herons that fed on the predatory fish had a concentration of 1000 units.
  a Give the term used to describe this change in concentration of pollutant in food chains.
  b Explain how the differences in mercury concentration happened.

9 A farmer decided to plough a hillside to plant corn and applied fertiliser to the crop in the spring. In summer there was a bloom of green algae in the lake below the hillside. By autumn fishermen complained that many of the fish had died.

  a Explain the link between applying fertiliser and the bloom of algae.

  b Suggest a reason why many of the fish died.

  c Give the appropriate term for these effects of nutrients in water.

  d Explain how farming breaks the rules of a closed-loop system, and how this makes it unsustainable.

10 Crop plants can be genetically modified to improve their qualities. For each statement below, explain if it is a risk or a benefit of GM crops.

  • Less of a GM crop is lost to pests and competition with weeds.

  • Added genes could produce unexpected toxins or allergens.

  • Added herbicide resistance means farmers can control weeds with special herbicides.

  • Added pest resistance means farmers don't need to use chemicals.

  • New crops could change ecosystems.

Explain your view of GM crops.

11 Bacteria and fungi can be used to make useful products by fermentation.

  a Describe a fermenter.

  b Explain what is needed for bacteria or fungi to grow well in a fermenter.

  c Name three different types of chemical made using bacteria and fungi.

12 This diagram shows genetic modification of a bacterium to produce human insulin.

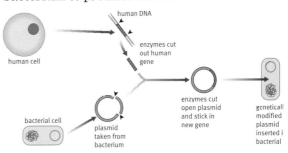

human DNA

enzymes cut out human gene

human cell

enzymes cut open plasmid and stick in new gene

bacterial cell

plasmid taken from bacterium

gcneticall modified plasmid inserted i bacterial

  a What is genetic modification?

  b In this diagram, what is the vector?

  c What does the vector do in this process?

  d How would you select modified bacteria?

  e What are the benefits of producing human insulin?

13 Chymosin is an enzyme produced commercially using genetically modified bacteria. It is used to make cheese from milk as an alternative to extracts from calves' stomachs.

  a What are the advantages of bacterial chymosin over calf chymosin?

  b Do you think it is right for manufacturers to use genetically modified bacteria in this way?

14 Stem cell technology could provide treatments for many medical conditions. Human embryos could provide a source of stem cells for treatments.

  a What is special about stem cells in the body?

  b How do healthy bone marrow stem cells help to treat leukaemia?

  c What are the concerns that some people have about how human embryos could be used?

Glossary

accuracy An accurate instrument or procedure gives a 'true' reading.

active site The part of an enzyme that the reacting molecules fit into.

active transport Molecules are moved in or out of a cell using energy. This process is used when transport needs to be faster than diffusion, and when molecules are being moved from a region where they are at low concentration to where they are at high concentration.

adaptation Features that help an organism survive in its environment.

ADH A hormone making kidney tubules more permeable to water, causing greater re-absorption of water.

aerobic respiration Respiration that uses oxygen.

alcohol The intoxicating chemical in wine, beer, and spirits. Causes changes in behaviour and may create long-term addiction.

algae Simple green water plants.

algal bloom Rapid growth of algae making the water green. It can be toxic.

allele Different versions of the same gene.

Alzheimer's disease A form of senile dementia caused by irreversible degeneration of the brain.

amino acids The small molecules that are joined in long chains to make proteins. All the proteins in living things are made from 20 different amino acids joined in different orders.

anaerobic Without oxygen.

anaerobic respiration Respiration that does not use oxygen.

antagonistic effectors Antagonistic effectors have opposite effects.

antagonistic pair Two muscles that work to move the same bone in opposite directions, for example, the biceps and triceps muscles.

antibiotic Drugs that kill or stop the growth of bacteria and fungi.

antibiotic resistant Microorganisms that are not killed by antibiotics.

antibiotics Drugs that kill or stop the growth of bacteria and fungi.

antibodies A group of proteins made by white blood cells to fight dangerous microorganisms. A different antibody is needed to fight each different type of microorganism. Antibodies bind to the surface of the microorganism, which triggers other white blood cells to digest them.

antigens Proteins on the surface of a cell. A cell's antigens are unique markers.

aorta The main artery that carries oxygenated blood away from the left ventricle of the heart.

aquaculture Farming in water, such as fish farming.

arrhythmia A problem with the heart in which the muscle does not contract regularly – the rhythm is lost.

arteries Blood vessels that carry blood away from the heart.

asexual reproduction When a new individual is produced from just one parent.

atrium (plural atria) One of the upper chambers in the heart. The two atria pump blood to the ventricles.

auxin A plant hormone that affects plant growth and development. For example, auxin stimulates growth of roots in cuttings.

axon A long, thin extension of the cytoplasm of a neuron. The axon carries electrical impulses very quickly.

bacteria Single-celled microorganisms that do not have a nucleus. Some bacteria may cause disease.

bacteriophage A type of virus that infects bacteria.

bacterium (plural bacteria) One type of single-celled microorganism. They do not have a nucleus. Some bacteria may cause disease.

base pairing The bases in a DNA molecule (A, C, G, T) always bond in the same way. A and T always bond together. C and G always bond together.

baseline data Data gathered at the start of a study or experiment so that patterns and trends can be established

behaviour Everything an organism does; its response to all the stimuli around it.

beta blockers Drugs that block the receptor sites for the hormone adrenaline. They inhibit the normal effects of adrenaline on the body.

bioaccumulation Build-up of chemicals in organisms as the chemicals travel through the food chain.

biodegradable Substances that can be broken down by microorganisms such as bacteria and fungi. Most paper and wood items are biodegradable, but most plastics are not.

biodiversity The great variety of living things, both within a species and between different species.

bioethanol Ethanol fuel produced by yeast fermentation of plant materials, such as cane sugar and sugar beet.

biofuels Fuel made from crops such as rape seed.

biogas Methane gas produced by the anaerobic digestion of organic material, such as farm animal manure.

blind trial A clinical trial in which the patient does not know whether they are taking the new drug, but their doctor does.

blood pressure The pressure exerted by blood pushing on the walls of a blood vessel.

body mass index Your body mass index is calculated using the formula BMI = body mass (kg) / [height (m)]^2, Tables will tell you if your body mass is healthy for your size.

bone Strong, rigid tissues making up the skeleton of vertebrates.

cancer A growth or tumour caused by abnormal and uncontrolled cell division.

capillary Tiny blood vessels that are one cell thick. They carry blood through the tissues between the arteries and veins.

capillary bed Large numbers of narrow blood vessels that pass through each organ in the body. Capillaries receive blood from arteries and return it to veins. Capillary walls are only one cell thick.

carbohydrate A natural chemical made of carbon, hydrogen, and oxygen. An example is glucose $C_6H_{12}O_6$. Carbohydrates include sugars, starch, and cellulose.

carbon cycle The cycling of the element carbon in the environment between the atmosphere, biosphere, hydrosphere, and lithosphere. The element exists in different compounds in these spheres. In the atmosphere it is mainly present as carbon dioxide.

carbon sink A system taking carbon dioxide from the air and storing it, for example, a growing forest.

carrier Someone who has the recessive allele for a characteristic or disease but who does not have the characteristic or disease itself.

cartilage Tough, flexible tissue found at the end of bones and in joints. It protects the end of bones from rubbing together and becoming damaged.

catalyst Chemical that starts or speeds up the rate of a reaction without being changed by it.

cause When there is evidence that changes in a factor produce a particular outcome, then the factor is said to cause the outcome, for example, increases in the pollen count cause increases in the incidence of hay fever.

cell The basic structural and functional unit of all living things.

cell membrane Thin layer surrounding the cytoplasm of a cell. It restricts the passage of substances into and out of the cell.

cell wall Rigid outer layer of plant cells and bacteria.

cellulose The chemical that makes up most of the fibre in food. The human body cannot digest cellulose.

central nervous system In mammals this is the brain and spinal cord.

cerebral cortex The highly folded outer region of the brain, concerned with conscious behaviour.

chlorophyll A green pigment found in chloroplasts. Chlorophyll absorbs energy from sunlight for photosynthesis.

chloroplast An organelle found in some plant cells where photosynthesis takes place.

chromosome Long, thin, threadlike structures in the nucleus of a cell made from a molecule of DNA. Chromosomes carry the genes.

chymosin Enzyme that breaks down proteins (a protease) found in calf stomachs. Fungi have been genetically modified to produce chymosin industrially for cheese-making.

clinical trial When a new drug is tested on humans to find out whether it is safe and whether it works.

clone A new cell or individual made by asexual reproduction. A clone has the same genes as its parent.

closed-loop system A system with no waste – everything is recycled.

cloud formation Evaporation of water, for example, from a forest, condensing into clouds.

combustion The process of burning a substance that reacts with oxygen to produce heat and light.

competition Different organisms that require the same resource, such as water, food, light, or space, must compete for the resource.

concentrated solution The concentration of a solution depends on how much dissolved chemical (solute) there is compared with the solvent. A concentrated solution contains a high level of solute to solvent.

conditioned reflex A reflex where the response is associated with a secondary stimulus, for example, a dog salivates when it hears a bell because it has associated the bell with food.

conditioning Reinforcement of behaviour associated with conditioned reflexes.

conscious To have awareness of surroundings and sensations.

consciousness The part of the human brain concerned with thought and decision making.

consumers Organisms that eat others in a food chain. This is all the organisms in a food chain except the producer(s).

control In a clinical trial, the control group is people taking the currently used drug. The effects of the new drug can then be compared to this group.

core (of the body) Central parts of the body where the body temperature is kept constant.

coronary artery Artery that supplies blood carrying oxygen and glucose directly to the muscle cells of the heart.

correlation A link between two things, for example, if an outcome happens when a factor is present, but not when it is absent, or if an outcome increases or decreases when a factor increases. For example, when pollen count increases hayfever cases also increase.

crop rotation Changing the crop grown in a field each year to preserve fertility.

crude oil Oil straight from an oil well, not refined into petrol or diesel.

cutting A shoot or leaf taken from a plant, to be grown into a new plant.

cystic fibrosis An inherited disorder. The disorder is caused by recessive alleles.

cytoplasm Gel enclosed by the cell membrane that contains the cell organelles such as mitochondria.

dead organic matter Any material that was once part of a living organism.

decomposer Organism that feeds on dead organisms. Decomposers break down the complex organic chemicals in their bodies, releasing nutrients back into the ecosystem to be used by other living organisms.

deforestation Cutting down and clearing forests leaving bare ground.

dehydration Drying out.

denatured A change in the usual nature of something. When enzymes are denatured by heat, their structure, including the shape of the active site, is altered.

denitrification Removal of nitrogen from soil. Bacteria break down nitrates in the soil, converting them back to nitrogen.

denitrifying bacteria Bacteria that break down nitrates in the soil, releasing nitrogen into the air.

deoxygenated Blood in which the haemoglobin is not bound to oxygen molecules.

desert Very dry area where no plants can grow. The area can be cold or hot.

desertification Turning to desert.

detritivore Organism that feeds on dead organisms and waste. Woodlice, earthworms, and millipedes are examples of detritivores.

development How an organism changes as it grows and matures. As a zygote develops, it forms more and more cells. These are organised into different tissues and organs.

diabetes type 1 An illness where the level of sugar in the blood cannot be controlled. Type 1 diabetes starts suddenly, often when people are young. Cells in the pancreas stop producing insulin. Treatment is by regular insulin injections

diabetes type 2 An illness where the level of sugar in the blood cannot be controlled. Type 2 diabetes develops in people with poor diets or who are obese. The cells in the body stop responding to insulin. Treatment is through careful diet and regular exercise.

diastolic The blood pressure when all parts of the heart muscle are relaxed and the heart is filling with blood.

differentiated A differentiated cell has a specialised form suited to its function. It cannot change into another kind of cell.

diffusion Movement of molecules from a region of high concentration to a region of lower concentration.

digest Break down larger, insoluble molecules into small, soluble molecules.

digestive enzyme Biological catalysts that break down food.

dilute The concentration of a solution depends on how much dissolved chemical (solute) there is compared with the solvent. A dilute solution contains a low level of solute to solvent.

dioxin Poisonous chemicals, for example, released when plastics burn.

direct drilling Planting seeds directly into the soil without ploughing first

disease A condition that impairs normal functioning of an organism's body, usually associated with particular signs and symptoms. It may be caused by an infection or by the dysfunction of internal organs.

dislocation An injury where a bone is forced out of its joint.

DNA (deoxyribonucleic acid) The chemical that makes up chromosomes. DNA carries the genetic code, which controls how an organism develops.

DNA fingerprinting A DNA fingerprint uses gene probes to identify particular sequences of DNA bases in a person's genetic make-up. The pattern produced in a DNA fingerprint can be used to identify family relationships.

DNA profiling A DNA profile is produced in the same way as a DNA fingerprint, but fewer gene probes are used. DNA profiling is used in forensic science to test samples of DNA left at crime scenes.

dominant Describes an allele that will show up in an organism even if a different allele of the gene is present. You only need to have one copy of a dominant allele to have the feature it produces.

double circulation A circulatory system where the blood passes through the heart twice for every complete circulation of the body.

double helix The shape of the DNA molecule, with two strands twisted together in a spiral.

double-blind trial A clinical trial in which neither the doctor nor the patient knows whether the patient is taking the new drug.

ecosystem Living organisms plus their non-living environment working together.

ecosystem services Life-support systems that we depend on for our survival.

Ecstasy A recreational drug that increases the concentration of serotonin at the synapses in the brain, giving pleasurable feelings. Long-term effects may include destruction of the synapses.

effector The part of a control system that brings about a change to the system.

embryo The earliest stage of development for an animal or plant. In humans the embryo stage lasts for the first two months.

embryo selection A process where an embryo's genes are checked before the embryo is put into the mother's womb. Only healthy embryos are chosen.

embryonic stem cell Unspecialised cell in the very early embryo that can divide to form any type of cell, or even a whole new individual. In human embryos the cells are identical and unspecialised up to the eight-cell stage.

endangered Species that are at risk of becoming extinct.

environment Everything that surrounds you. This is factors like the air and water, as well as other living things.

enzyme A protein that catalyses (speeds up) chemical reactions in living things.

epidemiological study Scientific study that examines the causes, spread, and control of a disease in a human population.

ethanol Waste product from anaerobic respiration in plants and yeast.

ethical Non-scientific, concerned with what is right or wrong.

ethics A set of principles that may show how to behave in a situation.

eutrophication Build-up of nutrients in water.

evolution The process by which species gradually change over time. Evolution can produce new species.

excretion The removal of waste products of chemical reactions from cells.

extinct A species is extinct when all the members of the species have died out.

fallow crop Crop that is not harvested, allowing the field to regain nutrients

false negative A wrong test result. The test result says that a person does not have a medical condition but this is incorrect.

false positive A wrong test result. The test result says that a person has a medical condition but this is incorrect.

fatty sheath Fat wrapped around the outside of an axon to insulate neurons from each other.

feral Untamed, wild.

fermentation Chemical reactions in living organisms that release energy from organic chemicals, such as yeast producing alcohol from the sugar in grapes.

fermenter Large vessel in which microorganisms are grown to make a useful product.

fertile An organism that can produce offspring.

fitness State of health and strength of the body

flowers Reproductive structures in plants often containing both male and female reproductive structures.

fluorescent marker Chemical attached to a DNA strand so it can be found or identified when separated from other strands in a gel.

food chain In the food industry this covers all the stages from where food grows, through harvesting, processing, preservation, and cooking to being eaten.

food web A series of linked food chains showing the feeding relationships in a habitat – 'what eats what'.

fossil The stony remains of an animal or plant that lived millions of years ago, or an imprint of its mark, for example, a footprint, in a surface.

fossil fuel Fuel made of the bodies of long-dead organisms.

fossil sunlight energy Sunlight energy stored as chemical energy in fossil fuel.

fruit Remaining parts of a flower containing seeds after fertilisation.

fungus (plural fungi) A group of living things, including some microorganisms, that cannot make their own food.

gametes The sex cells that fuse to form a zygote. In humans, the male gamete is the sperm and the female gamete is the egg.

gas exchange The exchange of oxygen and carbon dioxide that takes place in the lungs.

gene A section of DNA giving the instructions for a cell about how to make one kind of protein.

gene probe A short piece of single-stranded DNA used in a genetic test. The gene probe has complementary bases to the allele that is being tested for.

gene switching Genes in the nucleus of a cell switch off and are inactive when a cell becomes specialised. Only genes that the cell needs to carry out its particular job stay active.

genetic Factors that are affected by an organism's genes.

genetic modification (GM) Altering the characteristics of an organism by introducing the genes of another organism into its DNA.

genetic screening Testing a population for a particular allele.

genetic study Scientific study of the genes carried by people in a population to look for alleles that increase the risk of disease.

genetic variation Differences between individuals caused by differences in their genes. Gametes show genetic variation – they all have different genes.

genotype A description of the genes an organism has.

glands Parts of the body that make enzymes, hormones, and other secretions in the body, for example sweat glands.

glucose Sugar produced during photosynthesis.

habitat The place where an organism lives.

haemoglobin The protein molecule in red blood cells. Haemoglobin binds to oxygen and carries it around the body. It also gives blood its red colour.

heart disease A disease where the coronary arteries become increasingly blocked with fatty deposits, restricting the blood flow to the heart muscle. The risk of this is increased by a high fat diet, smoking, and drinking excess alcohol.

heavy metals Metals such as lead and mercury, which are toxic in small concentrations.

herbicide Chemical that kills plants, usually plants that are weeds in crops or gardens.

herbicide resistant Plants that are not killed by herbicides.

heterozygous An individual with two different alleles for a particular gene.

homozygous An individual with both alleles of a particular gene the same.

hormone A chemical messenger secreted by specialised cells in animals and plants. Hormones bring about changes in cells or tissues in different parts of the animal or plant.

human trial The stage of the trial process for a new drug where the drug is taken by healthy volunteers to see if it is safe, and then by sick volunteers to check that it works.

Huntington's disease An inherited disease of the nervous system. The symptoms do not show up until middle age.

hypothalamus The part of the brain that controls many different functions, for example, body temperature.

immune Able to react to an infection quickly, stopping the microorganisms before they can make you ill, usually because you've been exposed to them before.

immune system A group of organs and tissues in the body that fight infections.

infectious A disease that can be caught. The microorganism that causes it is passed from one person to another through the air, through water, or by touch.

infertile An organism that cannot produce offspring.

inherited A feature that is passed from parents to offspring by their genes.

innate Inborn, inherited from parents via genes.

insoluble Does not form a solution (dissolve) in water or other solutes.

insulin A hormone produced by the pancreas. It is a chemical that helps to control the level of sugar (glucose) in the blood.

intensive agriculture Farming with high inputs of fertiliser and pesticides and high productivity.

involuntary An automatic response made by the body without conscious thought.

joint A point where two or more bones meet.

kidney Organ in the body that removes waste urea from the blood, and balances water and blood plasma levels. People are usually born with two kidneys.

lactic acid Waste product from anaerobic respiration in animals.

learn To gain new knowledge or skills.

lichen Organism consisting of a fungus growing with a simple photosynthetic organism called an alga. Lichens grow very slowly are often found growing on walls and roofs.

life cycle The stages an organism goes through as it matures, develops, and reproduces.

lifestyle The way in which people choose to live their lives, for example, what they choose to eat, how much exercise they choose to do, how much stress they experience in their job.

lifestyle disease Disease that is not caused by microorganisms. They are triggered by other factors, for example, smoking, diet, and lack of exercise.

lifestyle history The way you have been living, taking regular exercise, eating healthily, and so on.

ligament Tissue that joins two or more bones together.

light intensity The amount of light reaching a given area.

light meter Device for measuring light intensity.

lignocellulase Enzyme that can break down the woody fibres in plant material (lignin) and the cellulose of plant cell walls.

limiting factor The factor that prevents the rate of photosynthesis from increasing at a particular time. This may be light intensity, temperature, carbon dioxide concentration, or water availability.

linear system A system based on the take-make-dump model.

lock-and-key model In chemical reactions catalysed by enzymes, molecules taking part in the reaction fit exactly into the enzyme's active site. The active site will not fit other molecules – it is specific. This is like a key fitting into a lock.

long-term memory The part of the memory that stores information for a long period, or permanently.

match Some studies into diseases compare two groups of people. People in each group are chosen to be as similar as possible (matched) so that the results can be fairly compared.

mayfly larvae Mayflies spend most of their lives (up to three years) as larvae (also called mayfly nymphs). They live and feed in aquatic environments. The adult insects live on the wing for a short time, from a few hours to a few days.

medical history Health or health problems in the past.

medication Any pharmaceutical drug used to treat or prevent an illness.

meiosis Cell division that halves the number of chromosomes to produce gametes. The four new cells are genetically different from each other and from the parent cell.

memory The storage and retrieval of information by the brain.

memory cell Long-lived white blood cell, which is able to respond very quickly (by producing antibodies to destroy the microorganism) when it meets a microorganism for the second time.

meristem cells Unspecialised cells in plants that can develop into any kind of specialised cell.

microorganism Living organism that can only be seen by looking through a microscope. They include bacteria, viruses, and fungi.

mitochondrion (plural mitochondria) An organelle in animal and plant cells where respiration takes place.

mitosis Cell division that makes two new cells identical to each other and to the parent cell.

models of memory Explanations for how memory is structured in the brain.

monoculture The continuous growing of one type of crop.

motor neuron A neuron that carries nerve impulses from the brain or spinal cord to an effector.

mRNA Messenger RNA, a chemical involved in making proteins in cells. The mRNA molecule is similar to DNA but single stranded. It carries the genetic code from the DNA molecule out of the nucleus into the cytoplasm.

multi-store model One explanation for how the human memory works.

muscles Muscles move parts of the skeleton for movement. There is also muscle tissue in other parts of the body, for example, in the walls of arteries.

mutation A change in the DNA of an organism. It alters a gene and may change the organism's characteristics.

nanometre A unit of measurement (abbreviation nm). A millimetre is the same as 1 million nanometres. $1 \text{ nm} = 1 \text{ m} \times 10^{-9} \text{ m}$)

native species Organisms naturally occurring in an area – not introduced by humans.

natural selection When certain individuals are better suited to their environment they are more likely to survive and breed, passing on their features to the next generation.

negative feedback A system where any change results in actions that reverse the original change.

nerve cell A cell in the nervous system that transmits electrical signals to allow communication within the body.

nerve impulses Electrical signals carried by neurons (nerve cells).

nervous system Tissues and organs that control the body's responses to stimuli. In a mammal it is made up of the central nervous system and peripheral nervous system.

neuron Nerve cell.

neuroscientist A scientist who studies how the brain and nerves function.

newborn reflexes Reflexes to particular stimuli that usually occur only for a short time in newborn babies.

nitrate ions An ion is an electrically charged atom or group of atoms. The nitrate ion has a negative charge, NO_3^-.

nitrogen cycle The continual cycling of nitrogen, which is one of the elements essential for life. By being converted to different chemical forms, nitrogen is able to pass between the atmosphere, lithosphere, hydrosphere, and biosphere.

nitrogen fixation When nitrogen in the air is converted into nitrates in the soil by bacteria.

nitrogen-fixing bacteria Bacteria found in the soil and in swellings (nodules) on the roots of some plants (legumes), such as clover and peas. These bacteria take in nitrogen gas and make nitrates, which plants can absorb and use to make proteins.

non-biodegradable Waste materials that microorganisms cannot break down.

nucleus Organelle that contains the chromosomes cells of plants, animals, fungi, and some microorganisms.

nucleus Central structure in a cell containing genetic material. It controls the function and characteristics of the cell.

obesity A medical condition where the increase in body fat poses a serious threat to health. A body mass index over 30 kg/m2.

open-label trial A clinical drug test in which both the patient and their doctor knows whether the patient is taking the new drug.

optimum temperature The temperature at which enzymes work fastest.

organ Part of a plant or animal made up of different tissues.

organelles The specialised parts of a cell, such as the nucleus and mitochondria. Chloroplasts are organelles that occur only in plant cells.

osmosis The diffusion of water across a partially permeable membrane.

overgrazing Too many grazing animals, such as goats, damaging the environment.

oxygenated Blood in which the haemoglobin is bound to oxygen molecules (oxyhaemoglobin).

pancreas An organ in the body which produces some hormones and digestive enzymes. The hormone insulin is made here.

partially permeable membrane A membrane that acts as a barrier to some molecules but allows others to diffuse through freely.

pathway A series of connected neurones that allow nerve impulses to travel along a particular route very quickly.

peripheral nervous system The network of nerves connecting the central nervous system to the rest of the body.

phagocytosis Engulfing and digestion of microorganisms and other foreign matter by white blood cells.

phenotype A description of the physical characteristics that an organism has (often related to a particular gene).

phloem A plant tissue that transports sugar throughout a plant.

photosynthesis The process in green plants that uses energy from sunlight to convert carbon dioxide and water into the sugar glucose.

phototropism The bending of growing plant shoots towards the light.

phytoplankton Single-celled photosynthetic organisms found in an ocean ecosystem.

pituitary gland Part of the human brain that coordinates many different functions, for example, release of ADH.

plasma The clear straw-coloured fluid part of blood.

plasmids Small circle of DNA found in bacteria. Plasmids are not part of a bacterium's main chromosome.

platelets Cell fragments found in blood. Platelets play a role in the clotting process.

pollen Plant reproductive structures containing a male gamete.

pollinators Animals, such as bees, that transfer pollen from anther to stigma.

polymer A material made up of very long molecules. The molecules are long chains of smaller molecules.

population   A group of animals or plants of the same species living in the same area.

predator   An animal that kills other animals (its prey) for food.

pre-implantation genetic diagnosis (PGD)   This is the technical term for embryo selection. Embryos fertilised outside the body are tested for genetic disorders. Only healthy embryos are put into the mother's uterus.

primary forest   A forest that has never been felled or logged.

processing centre   The part of a control system that receives and processes information from the receptor, and triggers action by the effectors.

producers   Organisms found at the start of a food chain. Producers are autotrophs, able to make their own food.

proportional   Two variables are proportional if there is a constant ratio between them.

protein   Chemicals in living things that are polymers made by joining together amino acids.

Prozac   A brand name for an antidepressant drug. It increases the concentration of serotonin at the synapses in the brain.

pulmonary artery   The artery that carries deoxygenated blood to the lungs. The artery leaves the right ventricle of the heart.

pulmonary vein   The vein that carries oxygenated blood from the lungs to the left atrium of the heart.

pulse rate   The rate at which the heart beats. The pulse is measured by pressing on an artery in the neck, wrist, or groin.

pupil reflex   The reaction of the muscles in the pupil to light. The pupil contracts in bright light and relaxes in dim light.

quadrat   A square grid of a known area that is used to survey plants in a location. Quadrats come in different sizes up to 1 m². The size of quadrat that is chosen depends on the size of the plants and also the area that needs to be surveyed.

quota   Agreed total amount that can be taken or harvested per year.

random   Of no predictable pattern.

rate of photosynthesis   Rate at which green plants convert carbon dioxide and water to glucose in the presence of light.

reactants   substances used in reactions by living organisms or by non living matter

receptor   The part of a control system that detects changes in the system and passes this information to the processing centre.

receptor molecule   A protein (often embedded in a cell membrane) that exactly fits with a specific molecule, bringing about a reaction in the cell.

recessive   An allele that will only show up in an organism when a dominant allele of the gene is not present. You must have two copies of a recessive allele to have the feature it produces.

recovery period   The time for you to recover after taking exercise and for your heart rate to return to its resting rate.

red blood cells   Blood cells containing haemoglobin, which binds to oxygen so that it can be carried around the body by the bloodstream.

reflex arc   A neuron pathway that brings about a reflex response. A reflex arc involves a sensory neuron, connecting neurons in the brain or spinal cord, and a motor neuron.

reject   How a body might react to foreign material introduced in a transplant.

relay neuron   A neuron that carries the impulses from the sensory neuron to the motor neuron.

reliability   How trustworthy data is.

rennet   An enzyme used in cheese-making.

repetition   Act of repeating.

repetition of information   Saying or writing the same thing several times.

reproductive isolation   Two populations are reproductively isolated if they are unable to breed with each other.

respiration   A series of chemical reactions in cells that release energy for the cell to use.

response   Action or behaviour that is caused by a stimulus.

retina   Light-sensitive layer at the back of the eye. The retina detects light by converting light into nerve impulses.

retrieval of information   Collecting information from a particular source.

ribosomes   Organelles in cells. Amino acids are joined together to form proteins in the ribosomes.

RICE   RICE stands for rest, ice, compression, elevation. This is the treatment for a sprain.

risk   A measure of the size of a potential danger. It is calculated by combining a measure of a hazard with the chance of it happening.

risk factor   A variable linked to an increased risk of disease. Risk factors are linked to disease but may not be the cause of the disease.

root hair cell   Microscopic cell that increases the surface area for absorption of minerals and water by plant roots.

rooting powder   A product used in gardening containing plant hormones. Rooting powder encourages a cutting to form roots.

sample   Small part of something that is likely to represent the whole.

selective breeding   Choosing parent organisms with certain characteristics and mating them to try to produce offspring that have these characteristics.

sensory neuron   A neuron that carries nerve impulses from a receptor to the brain or spinal cord.

serotonin   A chemical released at one type of synapse in the brain, resulting in feelings of pleasure.

sex cell   Cells produced by males and females for reproduction – sperm cells and egg cells. Sex cells carry a copy of the parent's genetic information. They join together at fertilisation.

sexual reproduction   Reproduction where the sex cells from two individuals fuse together to form an embryo.

shivering   Very quick muscle contractions. Releases more energy from muscle cells to raise body temperature.

short-term memory   The part of the memory that stores information for a short time.

silting of rivers   Eroded soil making the water muddy and settling on the river bed.

simple reflex   An automatic response made by an animal to a stimulus.

single-celled protein (SCP)   A microorganism grown as a source of food protein. Most single-celled protein is used in animal feed, but one type is used in food for humans.

skeleton   The bones that form a framework for the body. The skeleton supports and protects the internal organs, and provides a system of levers that allow the body to move. Some bones also make red blood cells.

social behaviour   Behaviour that takes place between members of the same species, including humans.

soil erosion   Soil removal by wind or rain into rivers or the sea.

specialised   A specialised cell is adapted for a particular job.

species   A group or organisms that can breed to produce fertile offspring.

sprain   An injury where ligaments are located.

stable ecosystem   An ecosystem that renews itself and does not change.

starch   A type of carbohydrate found in bread, potatoes, and rice. Plants produce starch to store the energy food they make by photosynthesis. Starch molecules are a long chain of glucose molecules.

starch grains   Microscopic granules of starch forming an energy store in plant cells.

stem cell   Unspecialised animal cell that can divide and develop into specialised cells.

stimulus   A change in the environment that causes a response.

stomata   Tiny holes in the underside of a leaf that allow carbon dioxide into the leaf and water and oxygen out of the leaf.

structural   Making up the structure (of a cell or organism).

structural proteins   Proteins that are used to build cells.

sustainability   Using resources and the environment to meet the needs of people today without damaging Earth or reducing the resources for people in the future.

sustainable   Able to continue over long periods of time.

symptom   What a person has when they have a particular illness, for example, a rash, high temperature, and sore throat.

synapse   A tiny gap between neurons that transmits nerve impulses from one neuron to another by means of chemicals diffusing across the gap.

synovial fluid   Fluid found in the cavity of a joint. The fluid lubricates and nourishes the joint, and prevents two bones from rubbing against each other.

tendon   Tissue that joins muscle to a bone.

termination   When medicine or surgical treatment is used to end a pregnancy.

therapeutic cloning   Growing new tissues and organs from cloning embryonic stem cells. The new tissues and organs are used to treat people who are ill or injured.

tissue   Group of specialised cells of the same type working together to do a particular job.

tissue fluid   Plasma that is forced out of the blood as it passes through a capillary network. Tissue fluid carries dissolved chemicals from the blood to cells.

torn ligament   An injury of the elastic tissues that hold bones together, a common sports injury of the knee. For treatment see 'RICE'.

torn tendon   An injury of the inelastic tissues that connect muscles to bones. For treatment see 'RICE'.

toxic   Poisonous.

transect   A straight line that runs through a location. Data on plant and animal distribution is recorded at regular intervals along the line.

transmitter substance   Chemical that bridges the gap between two neurons.

triplet code   A sequence of three bases coding for a particular amino acid in the genetic code.

unspecialised   Cells that have not yet developed into one particular type of cell.

vaccination   Introducing to the body a chemical (a vaccine) used to make a person immune to a disease. A vaccine contains weakened or dead microorganisms, or parts of the microorganism, so that the body makes antibodies to the disease without being ill.

valves   Flaps of tissue that act like one-way gates, only letting blood flow in one direction around the body. Valves are found in the heart and in veins.

variation   Differences between living organisms. This could be differences between species. There are also differences between members of a population from the same species.

vasoconstriction   Narrowing of blood vessels.

vasodilation   Widening of blood vessels.

vector   A method of transfer. Vectors are used to transfer genes from one organism to another.

vein   Blood vessel that carries blood towards the heart.

vena cava   The main vein that returns deoxygenated blood to the right atrium of the heart.

ventricle   One of the lower chambers of the heart. The right ventricle pumps blood to the lungs. The left ventricle pumps blood to the rest of the body.

virus   Microorganisms that can only live and reproduce inside living cells.

white blood cells   Cells in the blood that fight microorganisms. Some white blood cells digest invading microorganisms. Others produce antibodies.

working memory   The system in the brain responsible for holding and manipulating information needed to carry out tasks.

XX chromosomes   The pair of sex chromosomes found in a human female's body cells.

XY chromosomes   The pair of sex chromosomes found in a human male's body cells.

xylem   Plant tissue that transports water through a plant.

yeast   Single celled fungus used in brewing and baking.

yield   The crop yield is the amount of crop that can be grown per area of land.

zygote   The cell made when a sperm cell fertilises an egg cell in sexual reproduction.

Index

Appendices

Useful relationships, units, and data

Relationships

You will need to be able to carry out calculations using these mathematical relationships.

B7 Further biology

$$BMI = \frac{\text{body mass (kg)}}{[\text{height (m)}]^2}$$

Units that might be used in the Biology course

length: metres (m), kilometres (km), centimetres (cm), millimetres (mm), micrometres (µm), nanometres (nm)

mass: kilograms (kg), grams (g), milligrams (mg)

time: seconds (s), milliseconds (ms)

temperature: degrees Celsius (°C)

area: cm^2, m^2

volume: cm^3, dm^3, m^3, litres (l), millilitres (ml)

energy: joules (J), kilojoules (kJ), megajoules (MJ), kilowatt-hours (kWh), megawatt-hours (MWh)

power: watts (W), kilowatts (kW), megawatts (MW)

Prefixes for units

nano	micro	milli	kilo	mega	giga	tera
one thousand millionth	one millionth	one thousandth	× thousand	× million	× thousand million	× million million
0.000000001	0.000001	0.001	1000	1000 000	1000 000 000	1000 000 000 000
10^{-9}	10^{-6}	10^{-3}	$\times 10^3$	$\times 10^6$	$\times 10^9$	$\times 10^{12}$

Useful information

B4 The processes of life

photosynthesis
$$6CO_2 + 6H_2O \xrightarrow{\text{light energy}} C_6H_{12}O_6 + 6O_2$$

aerobic respiration $\quad C_6H_{12}O_6 + 6O_2 \longrightarrow 6CO_2 + 6H_2O$

OXFORD
UNIVERSITY PRESS

Great Clarendon Street, Oxford OX2 6DP

Oxford University Press is a department of the University of Oxford. It furthers the University's objective of excellence in research, scholarship, and education by publishing worldwide in

Oxford   New York

Auckland Cape Town Dar es Salaam Hong Kong Karachi
Kuala Lumpur Madrid Melbourne Mexico City Nairobi
New Delhi Shanghai Taipei Toronto

With offices in
Argentina Austria Brazil Chile Czech Republic France Greece Guatemala Hungary Italy Japan Poland Portugal Singapore
South Korea Switzerland Thailand Turkey Ukraine Vietnam

Oxford is a registered trade mark of Oxford University Press in the UK and in certain other countries.

© University of York and the Nuffield Foundation 2011.

The moral rights of the authors have been asserted.

Database right Oxford University Press (maker).

First published 2011.

British Library Cataloguing in Publication Data.

Data available.

ISBN 978-0-19-913832-6

10 9 8 7 6 5 4

Printed in Great Britain by Bell and Bain Ltd., Glasgow

Paper used in the production of this book is a natural, recyclable product made from wood grown in sustainable forests. The manufacturing process conforms to the environmental regulations of the country of origin.

Acknowledgements
The publisher and authors would like to thank the following for their permission to reproduce photographs and other copyright material:
P11t: Dr Jeremy Burgess/Science Photo Library; **P11m**: Dr Jeremy Burgess/Science Photo Library; **P11b**: Alexander Tsiaras/Science Photo Library; **P13l**: David Taylor/Science Photo Library; **P13r**: Martyn F. Chillmaid/Science Photo Library;**P14**: Alan Schein Photography/Corbis; **P16l**: Monkey Business Images/Shutterstock;**P16r**: Luca DiCecco/Alamy; **P17**: Richard J. Green/Science Photo Library; **P18**:Kenneth Sponsler/Shutterstock; **P19**: St. Felix School, Suffolk; **P21**: Mehau Kulyk/Science Photo Library; **P22l**: Dopamine/Science Photo Library;**P22r**: CNRI/Science Photo Library; **P23**: Oxford University Press; **P24l**: David Crausby/Alamy; **P24r**: BSIP ASTIER/Science Photo Library; **P25l**: Dan Sinclair/Zooid Pictures; **P25r**: Zooid Pictures; **P27**: Mauro Fermariello/Science Photo Library; **P28**: By Ian Miles-Flashpoint Pictures/Alamy; **P29**: BSIP, LAURENT/Science Photo Library; **P30**: Ariel Skelley/Corbis; **P31**: Mark Thomas/Science Photo Library; **P32**: Tek Image/Science Photo Library; **P35**: Jim Varney/Science Photo Library; **P36t**: BSIP, LAURENT H. AMERICAIN/Science Photo Library; **P36b**: Pascal Goetgheluck/Science Photo Library; **P37t**: David Scharf/Science Photo Library; **P37b**: Claude Nuridsany & Marie Perennou/Science Photo Library; **P38**: Dr Yorgos Nikas/Science Photo Library, **P39**: Phototake Inc./Alamy; **P42t**: BSIP, LAURENT/Science Photo Library; **P42b**: David Scharf/Science Photo Library; **P44**: Bjorn Svensson/Science Photo Library; **P46t**: Science Photo Library; **P46b**: Guzelian Photographers; **P47**: Guzelian Photographers; **P51**: Melanie Friend/Photofusion Picture Library/Alamy; **P53t**: Getty Images News/Getty Images; **P53b**: Philip Wolmuth/Alamy; **P54**: Robert Pickett/Corbis; **P55**: Paul A. Souders/Corbis; **P56t**: Donald R. Swartz/Shutterstock; **P56b**: Pete Saloutos/Corbis; **P57tl**: Simon Fraser/Mrc Unit, Newcastle General Hospital/Science Photo Library; **P57tr**: Ed Kashi/Corbis; **P57b**: Humphrey Evans/Cordaiy Photo Library Ltd./Corbis; **P58**: Dr P. Marazzi/Science Photo Library; **P60**: Guzelian Photographers; **P61t**: Science Photo Library; **P61b**: Biophoto Associates/Science Photo Library; **P62**: AVAVA/Shutterstock; **P63l**:

Bettmann/Corbis; **P63m**: Sipa Press/Rex Features;**P63r**: Matt Meadows, Peter Arnold Inc./Science Photo Library; **P65**: Janine Wiedel Photolibrary/Alamy; **P66**: Dimitri Iundt/TempSport/Corbis; **P67**: Publiphoto Diffusion/Science Photo Library; **P68**: Martyn F. Chillmaid; **P72**: Simon Fraser/Mrc Unit,Newcastle General Hospital/Science Photo Library; **P74**: Nancy Nehring/Istockphoto; **P76t**: Wayne Bennett/Corbis; **P76m**: Michael Prince/Corbis; **P76bl**: Kit Houghton/Corbis; **P76bm**: Corbis; **P76br**: Will & Deni McIntyre/Corbis; **P77t**: EuToch/Shutterstock; **P77b**: Stephen Ausmus/US Department Of Agriculture/Science Photo Library; **P78l**: Niall Benvie/Corbis; **P78r**: Alex Bartel/Science Photo Library; **P80t**: Alex Segre/Rex Features; **P80b**: Pakhnyushcha/Shutterstock; **P81l**: HartmutMorgenthal/Shutterstock; **P81r**: Dr Morley Read/Science Photo Library; **P83**: Wim Van Egmond, Visuals Unlimited/Science Photo Library; **P84**: Nigel Cattlin/Science Photo Library; **P85t**: Dr Keith Wheeler/Science Photo Library; **P85m**: Pedro Salaverria/Shutterstock; **P85b**: Duncan Shaw/Science Photo Library; **P87t**: Oxford University Press; **P87m**: Tom Brakefield/Corbis; **P87b**: Jeff Lepore/Science Photo Library; **P88t**: Holt Studios International; **P88b**: Steve Gschmeissner/Science Photo Library; **P96**: Konrad Wothe/LOOK-foto/Photolibrary; **P97**: source unknown (Georgina Mace); **P98t**: bl0ndie/Shutterstock; **P98m**: Collpicto/Science Photo Library; **P98b**: David Hartley/Rex Features; **P99t**: Paul Rapson/Science Photo Library; **P99b**: Serhiy Zavalnyuk/Istockphoto; **P102t**: bl0ndie/Shutterstock; **P102b**: Pakhnyushcha/Shutterstock; **P104**: Andrew Dorey/Istockphoto; **P107t**: Tfoxfoto/Shutterstock; **P107b**: James King Holmes/Science Photo Library; **P108**: J.C. Revy, Ism/Science Photo Library; **P109**: Eric and David Hosking/Corbis; **P110t**: Rick Price/Corbis; **P110b**: Simon Fraser/Science Photo Library; **P112**: J.C. Revy, Ism/Science Photo Library; **P113**: Dr Jeremy Burgess/Science Photo Library; **P114t**: Zooid Pictures; **P114b**: Manuel Blondeau/Photo & Co./Corbis; **P117t**: Biophoto Associates/Science Photo Library; **P117b**: Biophoto Associates/Science Photo Library; **P120**: Sinclair Stammers/Science Photo Library; **P121t**: Dr Jeremy Burgess/Science Photo Library; **P122**: P Phillips/Shutterstock; **P123**: Martyn F. Chillmaid/Science Photo Library; **P124**: K.R. Porter/Science Photo Library; **P125**: Kimimasa Mayama/Reuters; **P126l**: Aflo Foto Agency/Alamy; **P126r**: Blickwinkel/Alamy; **P127**: Power And Syred/Science Photo Library; **P128t**: Viktor1/Shutterstock; **128b**: Azpworldwide/Shutterstock; **P129l**: Ashley Cooper/Specialist Stock/Splashdown Direct/Rex Features; **P129r**: Eye Of Science/Science Photo Library; **P134**: Neil Bromhall/Science Photo Library; **P136tl**: Oxford Scientific Films/Photolibrary; **P136tm**: Oxford Scientific Films/Photolibrary; **P136tr**: Corel/Oxford University Press; **P136bl**: Michael & Patricia Fogden/Corbis; **P136bm**: Corel/Oxford University Press; **P136br**: Corel/Oxford University Press; **P138t**: Edelmann/Science Photo Library; **P138m**: Alexander Tsiaras/Science Photo Library; **P138b**: Science Photo Library; **P139t**: Bob Rowan; Progressive Image/Corbis; **P139b**: Leo Batten/Frank Lane Picture Agency/Corbis; **P140t**: Joe McDonald/Corbis; **P140b**: Astrid & Hanns-Frieder Michler/Science Photo Library; **P141t**: M J Higginson/Science Photo Library; **P141b**: Martyn F. Chillmaid/Science Photo Library; **P142l**: Oxford Scientific Films/photolibrary; **P142r**: John Kaprielian/Science Photo Library; **P145**: Rob Lewine/Corbis; **P147t**: Francis Leroy/Biocosmos/Science Photo Library; **P147b**: Corel/Oxford University Press; **P148tl**: Carl & Ann Purcell/Corbis; **P148tm**: Holt Studios International; **P148tr**: Martin Harvey/Corbis; **P148bl**: CNRI/Science Photo Library; **P148br**: Dr Jeremy Burgess/Science Photo Library; **P150t**: Science Photo Library; **P150b**: A.Barrington Brown/Science Photo Library; **P154l**: Bob Gibbons/Holt Studios International; **P154r**: Helen Mcardle/Science Photo Library; **P155tl**: Anthony Bannister/Gallo Images/Corbis; **P155tr**: Marko Modic/Corbis; **P155bl**: Russ Munn/Agstock/Science Photo Library; **P155br**: Foodfolio/Alamy; **P156t**: Biophoto Associates/Science Photo Library; **P156m**: Dr Yorgos Nikas/Science Photo Library; **P156b**: Dr. Y. Nikas/Phototake Inc/Alamy; **P157**: Edelmann/Science Photo Library; **P158**: Mauro Fermariello/Science Photo Library; **P159t**: Massimo Brega, The Lighthouse/Science Photo Library; **P159b**: Rex Features; **P162**: CNRI/Science Photo Library; **P164**: Sovereign, Ism/Science Photo Library; **P166tl**: Dennis Kunke/Phototake Inc./Alamy; **P166bl**: Astrid & Hanns-Frieder Michler/Science Photo Library; **P166br**: Eye Of Science/Science Photo Library; **P167tl**: Jeff Rotman/Nature Picture Library; **P167tr**: Tobias Bernhard/Oxford Scientific Films/Photolibrary; **P167bl**: Manfred Danegger/NHPA; **P167br**: Clive Druett/Papilio/Corbis; **P168l**: Sheila Terry/Science Photo Library; **P168m**: Owen Franken/Corbis; **P168r**: Laura Dwight/Corbis; **169l**: BSIP Astier/Science photo Library; **P169m**: Jennie Woodcock/Reflections Photolibrary/Corbis; **P169r**: Morgan McCauley/Corbis; **P173t**: Adam Hart-Davis/Science Photo Library; **P174t**: Greg Fiume/Corbis; **P174b**: Sipa Press/Rex Features; **P177t**: S. Kramer/Custom Medical Stock Photo/Science Photo Library; **P177b**: Mark Lythgoe and Steve Smith/Wellcome Trust; **P178**:Steve Bloom Images/Alamy; **P179**: Anthony Bannister/Gallo Images/Corbis; **P180t**: Lawrence Manning/Corbis; **P180m**: Kim Kulish/Corbis; **P180b**: Sally and Richard Greenhill/Alamy; **P183**: Jerry Wachter/Science Photo Library; **P184**: Karen Kasmauski/Corbis; **P185**: Roger Ressmeyer/Corbis; **P1888**: C. Warner Br/Everett/Rex Features; **P192**: Anthony Bannister/Gallo Images/Corbis; **P196**: Greg Epperson/Shutterstock; **P198tl**: Ingram Publishing/Photolibrary; **P198tr**: Juniors Bildarchiv/Photolibrary; **P198ml**: BSIP Medical/Photolibrary; **P198mr**: Dan Suzio/Science Photo Library; **P198bl**: Manfred Grebler/Photolibrary; **P198br**: Urban Zone/Alamy; **P199**: Firefly Productions/Corbis UK ltd.; **P202t**: Ian Hooton/Science Photo Library; **P202b**: DJ Mattaar/Shutterstock; **P203**: Jim DeLillo/Istockphoto; **P204l**: Widmann Widmann/Photolibrary; **P204r**: Adam van Bunnens/Alamy; **P207t**: Times Newspapers/Rex Features; **P207m**: Clive Brunskill/Getty Images Sports/Getty Images; **P208-209**: DJ Mattaar/Shutterstock; **P208**: Banana Stock/Photolibrary; **P211b**: Biophoto Associates/Science Photo Library; **P214r**: Klaus Guldbrandsen/

Science Photo Library; **P215:** Martin Dohrn/Royal College Of Surgeons/Science Photo Library; **P216:** Pasieka/Science Photo Library; **P218:** Dave Bartruff/ Corbis; **P219:** John Cleare Mountain Camera; **P220tl:** Lynne Harty Photography/Alamy; **P220tm:** Corbis UK Ltd.; **P220tr:** Wartenberg/Picture Press/Corbis; **P220bl:** Francesco Ridolfi/Shutterstock; **P220bm:** David Stoecklein/Corbis; **P220br:** Owen Franken/Corbis; **P221:** Mark Clarke/Science Photo Library; **P222:** Mark Clarke/Science Photo Library; **P223t:** Samuel Ashfield/Science Photo Library; **P223bl:** Steve Meddle/Rex Features, Alina555/ Istockphoto, Tomasz Zajaczkowski/Istockphoto, Charlotte Allen/Istockphoto, Elena Schweitzer/Istockphoto, Lauri Patterson/Istockphoto; **P223br:** Hulton Archive/Getty Images; **P224t:** Paul Harizan/Getty Images; **P224m:** Construction Photography/Corbis; **P224b:** ITV/Rex Features; **P226t:** Patrick Robert/Sygma/Corbis; **P226b:** Bettmann/Corbis; **P227t:** Janie Airey/Cancer Research UK; **P227b:** Moodboard/Corbis; **P228:** Ivonne Wierink/Shutterstock; **P229:** NatureOnline/Alamy; **P232t:** David M Dennis/Photolibrary; **P233:** Sergio/Pitamitz/Robert Harding/Rex Features; **P234t:** Mpemberton/ Dreamstime; **P234b:** Hugh Clark/FLPA; **P236tl:** S & D & K Maslowski/FLPA; **P236b:** Rodger Klein/Photolibrary; **P237l:** Ken Griffiths/NHPA; **P237r:** Mary Terriberry/Shutterstock; **P238:** Edward Webb/Rex Features; **P239:** No-till On The Plains; **P240:** Sean Gladwell/Shutterstock; **P242:** Tusia/Shutterstock; **P243t:** fazon1/Istockphoto; **P243b:** Dr Jeremy Burgess/Science Photo Library; **P244:** Thomas Winz/Photolibrary; **P245:** Moodboard RF/Photolibrary; **P246t:** Mark Edwards/Photolibrary; **P246b:** © RBG Kew; **P247tl:** Larissa Siebicke/ Photolibrary; **P247tr:** ImageBroker/Imagebroker/FLPA; **P247b:** Food and Agriculture Organisation of the United Nations; **P248:** Stasvolik/Dreamstime; **P249t:** Adrian Arbib/Photolibrary; **P249m:** Fletcher & Baylis/Science Photo Library; **P249b:** Francois Savigny/Nature Picture Library; **P251:** Conservation International; **P253:** Ashley Cooper, Visuals Unlimited/Science Photo Library; **P254t:** Sue Darlow/Photolibrary; **P254b:** Ruddy Gold/Photolibrary; **P264:** Jerry Mason/Science Photo Library; **P265:** Bill Barksdale/Agstockusa/Science Photo Library; **P267t:** T Philippe Plailly/Science Photo Library; **P267b:** Yoav Levy/ Photolibrary; **P268:** Eloy Alonso/Reuters; **P270t:** Steve Gschmeissner/Science Photo Library; **P270m:** Keys; **P270b:** Net Resources International; **P271:** Picsfive/Shutterstock; **P272t:** Pasieka/Science Photo Library; **P272b:** Mauro Fermariello/Science Photo Library; **P273:** Henrik Jonsson/Istockphoto; **P275tl:** Steve Allen/Science Photo Library; **P275tr:** BSIP, Raguet H/Science Photo Library; **P275bl:** David Leah/Science Photo Library; **P275br:** Hank Morgan/ Science Photo Library.

Illustrations by IFA Design, Plymouth, UK, Clive Goodyer, and Q2A Media.

The publisher and authors are grateful for permission to reprint the following copyright material:

Although we have made every effort to trace and contact all copyright holders before publication this has not been possible in all cases. If notified, the publisher will rectify any errors or omissions at the earliest opportunity.

Project Team acknowledgements
These resources have been developed to support teachers and students undertaking the OCR suite of specifications GCSE Science Twenty First Century Science. They have been developed from the 2006 edition of the resources. We would like to thank David Curnow and Alistair Moore and the examining team at OCR, who produced the specifications for the Twenty First Century Science course.

Authors and editors of the first edition
We thank the authors and editors of the first edition, Jenifer Burden, Anna Fullick, Anna Grayson, Angela Hall, Bill Indge, Pam Large, Jean Martin, Nick Owens, and Linn Winspear.
Many people from schools, colleges, universities, industry, and the professions contributed to the production of the first edition of these resources. We also acknowledge the invaluable contribution of the teachers and students in the pilot centres.
The first edition of Twenty First Century Science was developed with support from The Nuffield Foundation, The Salters Institute, and The Wellcome Trust. A full list of contributors can be found in the Teacher and Technician resources.

The continued development of *Twenty First Century Science* is made possible by generous support from:
- The Nuffield Foundation
- The Salters' Institute